Chico, George, the Birds, and Me

Number Forty-three:

Louise Lindsey Merrick Natural Environment Series

Chico, George, the Birds, and Me

THE MEXICAN TRAVELOGUE OF A WOMAN NATURALIST, 1948–1949

By Dorothy Chapman Saunders

Edited and annotated by
Henry M. Reeves and Roy E. Tomlinson

Epilogue by
Jesús G. Franco Pizaña

TEXAS A&M UNIVERSITY PRESS
College Station

First edition

This paper meets the requirements of ANSI/NISO Z39.48-1992
(Permanence of Paper).
Binding materials have been chosen for durability.

Library of Congress Cataloging-in-Publication Data

Saunders, Dorothy Chapman.
Chico, George, the birds, and me : the Mexican travelogue of a woman naturalist, 1948–1949 / by Dorothy Chapman Saunders ; edited and annotated by Henry M. Reeves and Roy E. Tomlinson ; epilogue by Jesús G. Franco Pizaña. — 1st ed.
p. cm. — (Louise Lindsey Merrick natural environment series ; no. 43)
Includes bibliographical references and index.
ISBN-13: 978-1-60344-061-5 (cloth : alk. paper)
ISBN-10: 1-60344-061-5 (cloth : alk. paper)
1. Saunders, Dorothy Chapman—Diaries. 2. Saunders, Dorothy Chapman—Travel—Mexico. 3. Naturalists—United States—Diaries. 4. Women naturalists—United States—Diaries. 5. Birds—Mexico. 6. Mexico—Description and travel. I. Reeves, Henry M. II. Tomlinson, Roy E. III. Title. IV. Series.
QL31.S18A3 2008
508.092—dc22
[B]
2008005212

To George,
who made it all
such fun.

Dorothy C. Saunders

Contents

George and Dorothy Saunders, marked "Colorado –'48 near Berthoud Pass," the year following their marriage and before they departed on their first Mexican trip. At the time they were living in Denver. This is the only photograph we have located of them together. (Source unknown)

Preface

During 1954–57, I (Reeves) was assigned to the Lower Rio Grande Valley of South Texas as a U.S. game management agent, U.S. Fish and Wildlife Service (FWS).

Fortunately, my office was situated adjoining that of Luther C. Goldman, manager of the Laguna Atascosa and Santa Ana National Wildlife Refuges. Unknowingly, Luther became my unofficial mentor. He was an excellent naturalist and superb field ornithologist, and he spoke Spanish. Further, both his father, Major Edward A. Goldman, and his uncle, Luther J. Goldman, had pioneered and conducted various early Bureau of Biological Survey (BBS) and, later, FWS biological investigations, including many in Mexico. As a youth, Luther C. sometimes accompanied them.

During our many discussions Luther sometimes spoke, always glowingly, of George B. Saunders, who years earlier had studied White-winged Doves in the Valley. One summer day in 1956, the legendary George B. Saunders appeared unexpectedly at the refuge office. Fortuitously, I happened to be present and we met. Toward the end of his visit, George asked if I would like to accompany him the following day as he revisited several White-winged Dove nesting colonies where he had conducted intensive studies in the late 1930s and early 1940s. I seized the opportunity. During the course of the day, however, George was devastated to see how many former brush tracts, some once accommodating more than a hundred nests per acre, had been cleared for agricultural use. Nonetheless, that day proved to be among the most memorable and educational during my FWS career.

I never saw George again. And I never had the opportunity to meet his wife, Dorothy C. Saunders, of whom I knew little until much later. However, while I was at the Patuxent Wildlife Research Center, Laurel, Maryland; later in Washington, D.C.; and after my retirement, George and I corresponded sporadically. Through correspondence, George entrusted me and fellow editor, Roy E. Tomlinson (also by correspondence), with various photographs and unpublished reports for which he no longer had file space.

In late summer of 1998 I received a large manila envelope from Dorothy. It contained a 331-page, double-spaced typed manuscript, on onionskin paper yellowed with age, titled "Two Go to See." Dorothy explained that she and George had been reviewing and disposing of old files when they uncovered her nearly forgotten manuscript written nearly a half century earlier. Her first inclination had been to discard it. But on second thought, and at George's urging, she decided to send it to me in the event that I might be interested in reading it.

Upon receiving the manuscript, I scanned through it, planning to set it aside for some future reading. However, tempted by the reading of a few random pages, I returned to its beginning. By the close of the following day, the manuscript had been read. It was a popular account of a five-month-long scientific investigation that she and George had made in Mexico during the winter of 1948–49. I thought, "What an entertaining, informative work it was, superb prose that should be preserved and, indeed, published."

Shortly after, I wrote Dorothy, telling her of my most favorable reaction and belief that the manuscript should be both preserved in at least two archives as well as published. She replied promptly, flattered by my words, saying that although she had not written it for publication, I could explore those possibilities for her. Copies were subsequently deposited, under the title "Two Go to See," in the Smithsonian Archives, Washington, D.C., alongside George's files, and in the Olin Library, Cornell University, Ithaca, New York. For various reasons, I was unable to pursue the publication possibility then, and the manuscript languished in my files. In the meantime, both George and Dorothy died.

Belatedly, in August 2004, the manuscript came to the forefront of things to do. I asked Roy Tomlinson, also a biologist retired from the FWS, and living in Albuquerque, New Mexico, if he would like to read Dorothy's treatise for general interest. Roy had conducted extensive research on migratory birds and endangered species in the U.S. Southwest and Mexico. We had worked together intermittently for nearly four decades, and I respected his ability as a researcher/editor/author. Roy agreed to read and evaluate Dorothy's story. Upon his completion of its reading, I was gratified when he independently concluded that the fascinating manuscript was worthy of publication. And I was especially delighted when Roy agreed to become a colleague, as coeditor, in pursuing that goal.

Our primary objective as coeditors of *Chico, George, the Birds, and Me* was to retain as much as possible of Dorothy's engaging, often poetic narration. Because Dorothy sometimes wandered a bit afield from her main themes, we exercised liberty in deleting some tangential passages. They generally included (1) repetitious biographical information; (2) extraneous accounts of travel other than in Mexico; (3) irrelevant shopping experiences; and (4) detailed accounts of merchandise manufacture. Transitional wording was inserted to bridge gaps caused by these deletions. Our corrections of the retained original text are minimal and confined to rare misspellings, punctuation errors, and the standardization of measurements. Endnotes address the very few factual errors we detected, as well as additions and annotations.

We attempted to hold our additions and annotations to a reasonable level while putting Dorothy's narration into a broader perspective. We also include pertinent references that provide the reader access to publications of special interest on Mexican subjects.

We sometimes clarified or expanded upon Dorothy's popular account with information from the Saunderses' other writing of their 1948–49 travels, specifically their unpublished official "Report on Migratory Waterfowl Investigations in Mexico, January–May 1949" (1950) and the published *Waterfowl and Their Wintering Grounds in Mexico, 1937–64* (1981).[1]

Overall, we estimate that approximately 85 percent of Dorothy's 331-page manuscript appears largely verbatim in this book. Inasmuch as Dorothy expressly did not write her personal account for publication, the editors assume full responsibility for errors that may be detected. Roy and I hope that readers will experience the same appreciation and enjoyment that we did upon our first readings. We believe that *Chico, George, the Birds, and Me* is an intriguing account of wildlife and their habitats in Mexico shortly following World War II. It also colorfully describes rural Mexico and its inhabitants at that formative time in our neighbor's history. Further, it is a testament to the loving partnership and lives of an extraordinary couple whose close relationship and cooperative professional achievements are found so seldom today. We hope that this book will bring a better recognition to Dorothy Chapman Saunders, a uniquely gifted woman. However, we are saddened that Dorothy's death precluded her knowing that her wonderful account was in fact published, albeit belatedly.

Acknowledgments

We greatly appreciate the varied forms of assistance given us by several retired U.S. Fish and Wildlife Service colleagues who personally knew Dorothy's husband, George, and some, indirectly of her. They include Luther C. Goldman (deceased), College Park, Maryland; Paul F. Springer (deceased), Arcata, California; Arthur S. Hawkins (deceased), Hugo, Minnesota; Walter F. Crissey, Palm Beach Gardens, Florida; and William H. Kiel, Kingsville, Texas. Richard C. Banks and Craig Ludwig, both of the Bird Section, National Museum of Natural History, Washington, D.C., provided information on birds collected by the Saunderses in Mexico for research purposes. Rod C. Drewien, of Vale, Oregon, helpfully provided unpublished technical reports and other information about early Mexican waterfowl surveys conducted by the Saunderses and others.

During his later years, George sporadically distributed reports, memos, and especially photographs and color slides to those with whom he corresponded (Drewien, Tomlinson, Reeves, and David Dolton, FWS, Denver, Colorado). The bulk of the illustrations appearing in the book are from these sources with additional ones provided by Tomlinson. Gary L. Kramer, Willows, California, another FWS retiree, supplied several helpful references. We especially thank Mercedes Foster, Smithsonian Institution, Washington, D.C.; and Jesús Franco Pizaña, Texas Parks and Wildlife Department, McAllen, Texas, for their helpful reviews of the entire manuscript. We thank Judith L. Villanova, Dorothy's trustee, Vero Beach, Florida, for information on her death. Of all the above, only she knew Dorothy personally. James Teer, Welder Wildlife Foundation (retired), read an early draft of the book and brought it to the attention of the Texas A&M University Press, for which we are grateful. Finally, we thank copyeditor Cynthia Lindlof and the many members of the Texas A&M University Press staff, particularly Shannon Davies and Jennifer Ann Hobson, for their careful and dedicated efforts to bring about the publication of *Chico, George, the Birds, and Me.*

To all, we are indebted.

Introduction

CHICO, GEORGE, *the Birds, and Me* is Dorothy Saunders's nearly daily personal account of a five-month-long ornithological expedition that she and her husband, George, made in Mexico during the winter of 1948–49. George, a research biologist of the Bureau of Biological Survey (BBS), U.S. Department of Agriculture, had been instructed to conduct field surveys that winter of waterfowl and wetlands over much of southern Mexico. Ancillary research was to be directed to continuing his studies of White-winged Doves and to avian wildlife in general. Dorothy was appointed as a field cooperator (unpaid) to make botanical observations and collections and to assist George. Her original manuscript and this edited book version are her personal accounts of their travels. She said that she had written it solely for personal refreshment of the wonderful days that she and George shared on their first extended field trip to Mexico. Shortly before her death, she graciously invited us to accompany them, as readers, and share in their fascination with our southerly neighbor, its wildlife, people, culture, and history.

What do we know of the remarkably gifted Dorothy Chapman Saunders and of her husband-scientist, George B. Saunders? Since readers will intimately accompany them for five months in the field in Mexico, it seems fitting, therefore, that readers know a bit about their companions.

Dorothy Chapman Saunders

Dorothy Chapman was born on April 2, 1912, in Hackensack, New Jersey. As a precocious child, the daughter of a New England Protestant minister-turned-banker father and a former school vice-principal mother, Dorothy possessed extraordinarily broad interests and talents that would lead her to exotic places and into endeavors foreign to women of her time. She undertook her undergraduate studies in journalism at Syracuse University, Utica, New York, and graduated with an A.B. (cum laude, Phi Beta Kappa) in 1934. She acquired an M.S. in 1935 at the University of Michigan, Ann Arbor, and a Ph.D. (botany), with high honors, in 1937 at the same university.[1]

Somehow along her academic way she earned a pilot's license, bought and flew an airplane, was a newspaper reporter, participated in sports (including baseball, basketball, swimming, skiing, and rifle marksmanship), was a ham-radio operator (she once talked with Admiral Richard E. Byrd in Antarctica), learned to play the piano and the recorder (the wooden medieval instrument, not the electronic device of today), traveled widely, and learned French, Latin, and Spanish.

Following graduation from Michigan, she served as professor of biology at Cedar Crest College, Allentown, Pennsylvania, during 1937–41. During 1941–44 she lived in Lima, Peru, where she studied art at the Escuela Nacional de Bellas Artes and the Universidad de San Marcos. She worked during 1944–47 as a research agriculturalist for the Office of Foreign Agricultural Relations of the U.S. Department of Agriculture, with extended assignments in Peru, Ecuador, El Salvador, Cuba, and Guatemala. Her work focused on agricultural products such as sugar, coffee, quinine, vanilla, bananas, and kenaf, all essential to the war effort. Her research eventually carried her to eighteen Central and South American countries, plus Mexico. It was at the U.S. embassy in Guatemala that she met George Saunders, whom she married in June 1947. Thereafter she worked as his unpaid field assistant, with the position title of scientific collaborator (research), U.S. Fish and Wildlife Service.

Dorothy's professional diversions included a postdoctoral Fulbright research scholarship to study blood parasites of fishes in the Red Sea, ichthyological research in Bermuda and the Bimini Islands for the National Science Foundation, teaching as an assistant professor of biology at the University of Florida during 1953–54, and conducting shark research at the Cape Haze Marine Laboratory in Sarasota, Florida, during 1955–73. Concomitant avocations encompassed painting and sketching, music, photography, collecting archaeological relics and works of art, genealogy, reading, and cooking and enjoying fine foods and life's other pleasures. She obviously had a keen zest for life, and she met every day with anticipation, enthusiasm, wonder, and joy.

Dorothy Saunders is listed in *American Men and Women of Science*, the second edition of *Who's Who in American Women*, and in the first edition of the *World's Who's Who in Science*. She published over forty scientific articles and two books on genealogy during her long, varied, and distinguished career. Dorothy Saunders was a Renaissance woman if ever there was one.[2]

She died on January 26, 2002, in Vero Beach, Florida, leaving no survivors. Her death followed George's by less than a year.

George B. Saunders

George Bradford Saunders Jr., born in Oklahoma City, Oklahoma, on June 8, 1907, was among the "third generation" of field biologists of the U.S. Fish and Wildlife Service and its forerunner, the Bureau of Biological Survey. Over his professional career, George became renowned for his studies of waterfowl and their habitats in Canada and Mexico (Saunders 1952, 1964), his rangewide studies of White-winged Doves (Saunders 1968), biological reconnaissance leading to the creation of several national wildlife refuges (notably Aransas, Laguna Atascosa, and Santa Ana, all in Texas), and for his general lifelong interest in all forms of wildlife.

Saunders earned his Ph.D. in 1932 from Cornell University, Ithaca, New York, for his taxonomic studies of the meadowlarks, genus *Sturnella*. After making an ornithological expedition to southern Africa for the Philadelphia Academy of Sciences, and serving as a research fellow for the National Research Council, he was appointed the staff state ornithologist for Michigan. In 1937 he joined the federal BBS, U.S. Department of Agriculture, as that agency's first Central Flyway biologist, working under the direction of Frederick C. Lincoln, founder of the avian flyway concept.

In his capacity as a flyway biologist, George literally followed waterfowl from their Canadian and U.S. breeding grounds into Mexico and Central America. He was among the first waterfowl biologists to champion the use of low-flying aircraft to survey waterfowl and their habitats. After 1939 his primary summer assignments were to study the White-winged Dove, an important game bird in South Texas and the Southwest, whose local abundance was sometimes likened to that of the passenger pigeon. For thirty-seven years George conducted field studies of migrant birds over widely diverse areas of North America from residences in Denver, Colorado; Brownsville, Texas; and Atlanta, Georgia.

Shortly after World War II, George led a team of U.S. biologists assigned

George B. Saunders (right) and David L. Spencer (pilot) at the Brownsville, Texas, airport following their completion of an aerial survey of wintering waterfowl in Mexico and Central America. The Fish and Wildlife Service's amphibian Grumman *Widgeon*, notably underpowered for such work, was replaced by the more powerful Grumman *Goose*. George was a strong advocate for the use of aircraft to survey waterfowl. (Photo by Luther C. Goldman)

the task of inventorying the wildlife of Guatemala.[3] Dorothy accompanied and collaborated with George on many productive and exciting trips afield in Latin America. Their loving relationship endured for fifty-three years. George died on February 15, 2001, at age ninety-three in Vero Beach, Florida, where he and Dorothy had spent their busy retirement years (for his obituary, see Reeves 2001; see also Saunders 1984).

Saunders was senior coauthor of the book *Whitewings, the Life History, Status, and Management of the White-winged Dove* (1968) and several scientific articles and many field reports, which are filed in the Smithsonian Institution Archives, Washington, D.C., as the George B. Saunders Papers, 1939–1993, Accession 99-074. George was also noted in *American Men and Women of Science.* Bibliographical citations of both Dorothy and George Saunders can be found in the references section.

Saunders discussed his career in *Flyways* (Hawkins et al. 1984) in which he lovingly wrote, "Judging from my own career, I believe that wildlife biologists are among the happiest people to be found, especially if they have a wife who is also a biologist and shares their adventures in the wild" (129).

The U.S.-Mexico Bird Treaty

A bit of further background sets the stage for the Saunderses' 1948–49 Mexican expedition and Dorothy's account of it.

In 1909, after hearing reports of the unregulated killing of wildlife in Mexico, particularly migratory waterfowl, T. Gilbert Pearson of the National Association of Audubon Societies traveled there to obtain firsthand information. He saw waterfowl and other birds being sold in the markets and streets of Mexico City and heard of the vast slaughters of waterfowl in the Valley of Mexico by armadas.[4] Pearson met with President Don Porfirio Díaz and suggested that Mexico consider legislation for protecting birds and game species, setting hunting seasons, establishing a hunting license, and creating a national wildlife protection agency.

Sporadic diplomatic progress by both the United States and Mexico foundered for various reasons, and not until early 1936 did both governments seriously turn to formally adopting a treaty similar to the one with Canada. The latter treaty, concluded in 1916, and its implementing legislation, the Migratory Bird Treaty Act (1918), designated which birds were to be protected and how hunting and other uses of certain species were to be regulated. The U.S.-Mexican treaty, which protected certain mammals as well as birds, gave further impetus to cooperative efforts of the two nations to protect certain shared species.[5] Although Dorothy Saunders does not mention it, the obligations of the Migratory Bird Treaty Act and the U.S.-Mexican treaty were the main justifications for their being in Mexico conducting field studies of waterfowl and their habitats.[6]

The government of Mexico for years had only a small agency, the Dirección General de Fauna Silvestre, based in Mexico City, to conduct research and management operations for the entire country. In the past two to three decades, however, the federal and state governments have established larger and better-trained and better-equipped staffs to handle a multitude of wildlife management and research programs. Although biologists from the United States still provide some guidance through the Tri-partite Joint Agreement on Wildlife Management involving Canada, Mexico, and the United States, Mexico now functions as an equal partner in migratory bird conservation.

Challenges to Conducting Fieldwork in Mexico

One cannot comprehend the accomplishments of the Saunderses during their 1948–49 investigations without knowing of the practical problems they encountered and overcame. Their biggest asset was that both spoke Spanish. Nonetheless, their obstacles were indeed formidable.

Their first impediment was simply gaining admittance into Mexico. Separate permits had to be obtained for themselves (visas), their vehicle, and guns for collecting specimens.[7] The process could take hours—even days.

A major difficulty was the poor state of roads in the rural areas they traveled, and to reach some remote areas, they resorted to trails or simply drove cross-country. Related to the road problem was the general unavailability and inadequacy of detailed maps of Mexico, especially of remote areas.[8] Road travel, particularly at night, was especially dangerous because of wandering livestock.[9] We speculate that the Saunderses had to carry extra cans of gasoline, especially in remote areas.

Probably their biggest worry was the performance of their "brand-new" jeep truck, Chico. It was underpowered because of engine problems and for most of their trip did not generate speeds over thirty-five miles per hour. Most towns did not have regular garages and repair services. So, they had to rely on self-taught mechanics at local repair shops. Fortunately, engines were much less complicated then, and George usually was able to keep the jeep running until they reached a trained mechanic.

Because they camped out much of the time, a host of impediments impacted what could be described as "human needs and comforts." Safe food and water were top priority.[10] Neither could be depended upon as being available and safe, even in many towns. The Saunderses had to make sure that they were served *agua purificada* in hotels and restaurants. Mosquitoes, chiggers, ticks, and other insects posed potential health problems as well as immediate discomforts. Only occasionally does Dorothy complain of excessive heat, wind, rain, and humidity, and never of the absence of air conditioning. Certainly, she was a stoic.

Physical safety was a matter of real concern in at least one instance when they were threatened by pulque-drunken fishermen.[11] However, by exercising reasonable precautions, they traveled widely and safely, perhaps even more so than we can in our own country today. Likewise, the few instances of dishonesty they encountered can be found anywhere. As trying and dangerous as Mexico could be, it was exciting and exotic. Both Dorothy and George clearly found it compelling and rewarding.

A bit of further background sets the stage for the Saunderses 1948–49 Mexican expedition and Dorothy's account of it. Also, a few other things mentioned in Dorothy's account require some further elucidation.

Specimen Collection

The National Museum of Natural History in Washington, D.C., requested George to collect biological specimens, particularly birds, during his Mexican investigations. Thus he, and even Dorothy, selectively killed many birds that George prepared as study skins for the museum. This may be thought of by some as needless slaughter; however, for scientific purposes such as taxonomy and classification, bird morphology and plumage had to be examined closely, something often impossible to do in the field. The development of better binoculars and spotting scopes, improvement of film and digital cameras and lenses, and the use of DNA expertise lessens the need today for collecting birds by shooting.

Correspondence with curators at the museum revealed that George Saunders is credited with providing 212 bird specimens from the 1948–49 trip. Additionally, Dorothy contributed an undetermined number of pressed and dried plant specimens.

Sale of Wild Birds

Wild birds were commonly sold in Mexico in the 1940s, and this practice probably continues in many sections to this day, although laws have been enacted to prohibit such activity. Nearly every *mercado* (town market) contained one shop that sold birds, usually surreptitiously. One of the editors (Tomlinson) has observed poachers using various methods to capture birds, including trapping and robbing nests of young. Brightly colored parrots, parakeets, and orioles are the most sought-after species, but certain beautiful songsters such as mockingbirds, solitaires, and White-winged Doves were also captured. This practice has led to the near extinction of various species, particularly the psittacines (parakeets, macaws, and parrots) in some locales. Surprisingly, Dorothy did not express much alarm at the practice that so obviously could affect local bird populations.

Dorothy's Ornithological Expertise

Although later in life Dorothy became a very proficient birder or bird watcher, we think that she was a relative novice in 1948–49. She was a capable botanist, but relatively new to ornithology as compared to her botanical competence. Nonetheless, she writes of seeing more than 170 species, a few of which she describes in some detail. Whenever a bird is first mentioned throughout the text and its name is incorrect according to the present American Ornithologists' Union *Check-list,* the correct name has been inserted in brackets. See the appendix for a listing of all birds mentioned in the text, including their scientific and Mexican names.

With this background setting the stage, we move on to Dorothy Saunders's *Chico, George, the Birds, and Me.*

Chico, George, the Birds, and Me

Sierra San Pedro Martir
BAJA CALIFORNIA DEL NORTE
Baja California Desert
BAJA CALIFORNIA DE SUR
Sierra San Lazaro
SONORA
Sonoran Desert
CHIHUAHUA
COAHUILA
NUEVO LEÓN
McAllen
Reynosa
Monterrey
Montemorales
Sierra Madre Oriental
TAMAULIPAS
Ciudad Victoria
Ciudad Mante
Tampico
Sierra Madre Occidental
SINALOA
Pacific Coastal Plain
Mazatlán
DURANGO
Central Plateau
ZACATECAS
SAN LUIS POTOSÍ
Ciudad Valles
Tamazunchale
Gulf Coastal Plain
NAYARIT
Tepic
AGUASCALIENTES
GUANAJUATO
Bajío
QUERÉTARO
HIDALGO
Guadalajara
JALISCO
Laguna de Chapala
Zamora de Hidalgo
Laguna de Cuitzéo
Morelia
Volcanic Cordillera
Laguna de Pátzcuaro
Uruapan
COLIMA
Colima
MICHOACÁN
San Blas Basin
Toluca
México City
MÉXICO
TLAXCALA
Xalapa
Veracruz
Popocatépetl
Cuernavaca
Puebla
Córdoba
Orizaba
MORELOS
Tehuitzing
PUEBLA
VERACRUZ
GUERRERO
Tierra Colorado
Acapulco
Laguna Papagayo
OAXACA
Sierra Madre de Oaxaca
Oaxaca
Monte Albán
Sierra Madre del Sur
Salina Cruz
Tehuantepec
Isthmus de Tehuantepec
TABASCO
CAMPECHE
YUCATAN
Yucatán Peninsula
QUINTANA ROO
Tuxtla Gutiérrez
San Cristóbal
Chiapas Highlands
CHIAPAS
Sierra Madre de Chiapas
Chiapas Coastal Plain

CHAPTER I

To Mexico with George, Chico, and Chiquita

December 22, Monterrey, Nuevo León

WE FINALLY crossed the border from Texas at Reynosa today and are in Mexico with all our goods and chattels intact. There have been months of preparation on the part of the U.S. Fish and Wildlife Service office in Washington, D.C., and the American Embassy in Mexico City, D.F., to get the necessary permits and papers from the Mexican government. And considerable preparations on our part by each having half a dozen inoculations, and gathering and packing all of our needed equipment for five months in the country. But for a time this morning it looked as though we might not be in Mexico tonight after all.

No matter how many papers one has, it always seems that the local Mexican customs officials require another which inevitably one hasn't. They are polite about it, but always firm. Today by some miracle we did not learn that we needed it five minutes before they closed their office for the long noon siesta. Or five minutes before they shut up shop for the day. We had time, by dint of considerable agility and good luck, and various trips between the two Mexican customs offices, which are half a mile apart, and the U.S. Consulate, to fill in the necessary forms and collect the needed signatures by noon.

By twelve-thirty P.M. we were through the town and merrily bowling along the highway to Monterrey. Perhaps, though, bowling isn't quite the right term. Our new three-quarter-ton jeep truck still apparently is making about half speed, although we have been in two garages to have the governor disconnected and the carburetor adjustments made. So our top speed on the level is in the neighborhood of thirty mph and an upgrade of any sort slows us down to low gear. However, we don't really care as yet. We are both going to see much of Mexico this winter that we haven't seen before, the progress of field studies sounds wonderful, and we're both well and happy.

My husband, George, who is a flyway research biologist with the U.S. Fish and Wildlife Service, will be studying and mapping the wintering grounds of waterfowl plus various phases of duck and other game-bird migrations between the United States and Mexico.[1] I am to be a "Scientific Collaborator (Research)," which means mostly, I am sure, dredging up and drying unsavory muddy masses of water plants which the waterfowl may eat.[2] I also suspect that I will serve as first assistant helper in everything from collecting and studying the plant contents of birds' stomachs to filling our own, and keeping a running road log of the important things we see as we jog along the highways—and much in between.

The countryside is so arid and barren between Reynosa and Monterrey that we saw very few birds or other animals today. However, among those few were three new ones for my life list of birds seen—the curve-billed thrasher, the Texas sparrow [Olive Sparrow], and the white-necked raven [Chihuahuan Raven]. (See Appendix for bird list.)

The latter seems very unfairly named to me from the viewpoint of an amateur trying to identify it in the field. I looked carefully through binoculars at the flock of them and saw not a single white feather on the necks, or anywhere else, could I see. George says that on a museum specimen one can clearly see that the bases of the neck feathers are white. But that is certainly no help to anyone identifying the much alive birds out of one's reach. They look entirely black like any other crow or raven.

Monterrey, which means "King's Mountain" in Spanish, is beautifully situated at a little over eighteen hundred feet in altitude in the valley of the Santa Catarina River. On three sides it is encircled by jagged mountains, which add greatly to its natural charm. Because it is so near the border, it is very much Americanized, too much for my taste. But of course on the credit side it is clean, progressive, and interested in and busy with manufacturing, schools, etc.

It was first heard of in the history about the year 1560 when the Spanish conquistadores enlarged the scope of their activities to include the northern part of Mexico. At that time it was founded as a little outpost village with the name Santa Lucía. In 1585 the name was changed to Ciudad de León (City of the Lion). And in 1596 it became La Ciudad Metropolitana de Nuestra Señora de Monterrey, the "Metropolitan City of Our Lady of Monterrey," now mercifully shortened to the last word.

Since local marketplaces are always high points of interest to me, we walked over to visit the one here after dinner.[3] Besides the usual foodstuffs there were quantities of freshly shelled pecans. We bought five pounds as we like to eat them along with dates, chocolate, and raisins when working long days in the field where there isn't either time or a place to prepare a regular lunch.

We were also delighted to see cellophane-wrapped bricks of the concentrated fruit pastes called *dulces* or *pastas*, especial favorites of George's. They are sliced and commonly served as dessert in many of the smaller restaurants and hotels throughout the country. And they keep wonderfully and so provide something to calm an aching sweet tooth when in hot areas where chocolate remains in a solid state for only a matter of minutes. We bought several, one each of guava, strawberry, quince, and pear, and my dessert problem while camping out is now considerably simplified.

Then there were fine handmade leather goods of all sorts, and rows of short-sleeved blouses embroidered with brilliantly colored flowers. One with crimson roses, each as large as a salad plate and with two-toned green leaves, caught my eye. It is very much like the design and work that the Indio women in Yucatán use for their own longer white *huipiles*.[4] Since it seemed the right size I bought it, and it should be just the thing for hot summer afternoons or evenings in Denver.

After the terrific heat of the drive from Reynosa today, the cool evening air is the most welcome thing we can think of. And we are spending the evening sitting outside our cottage in deck chairs soaking it all up, a long cold drink in hand, before turning in for the night.

December 23, Ciudad Mante, Tamaulipas

Our first pressing problem today has been the naming of the jeep. Most Mexican truck drivers regard their wheeled vehicles very much as a part of the family. And even the wheelbarrow of a road worker is liable to be painted by its proud owner with a name or philosophic title. The big truck trailers know no bounds in grandiose labels and flourishes of cartography on their bumpers and cabs. Two of the first we saw on the road today bore *El Sagrado Burrito,* "The Sacred Little Burro," and *Así es la Vida,* "Thus is Life," and started us thinking about the jeep.

Because of its junior size and short stubby nose and the ease of remembering and saying the word, we have chosen Chico, "Little One" in Spanish, for its name. Although for some hours I held out for Chatito, "Little Snub Nose," which I still think is a good name too. As a logical extension of the idea, that gives us Chiquita for the slender sixteen-foot, khaki-colored aluminum canoe we carry inverted and strapped on top of Chico. A boat is traditionally feminine. And it is only common politeness to use the Spanish affectionate diminutive when speaking of something we hope will cooperate in carrying us over many miles of water.

We drove out from Monterrey this morning past the spectacularly notched mountain peak known as La Silla, "The Saddle." Almost all day we have been going through semi-arid country, alternating with the green patches of citrus orchards. Off to our right, the rugged sweep of the great Sierra Madre Oriental mountain range parallels our route. Typical Mexican mountains are greatly broken and sweep down in great violet and purple folds and curves, the mountains of a geologically young country. Aside from the northern Gulf Coast and the Yucatán Peninsula, it is difficult to be anywhere in Mexico beyond the sight of at least one mountain. And usually behind the first one there are rows of others, each a little higher and a little fainter in tint than that before it.

I never fail to expect the Monterrey–Mexico City stretch to be like the beautiful highlands of Guatemala, densely forested in many localities and greatly broken up into mountains and valleys. It never seems fitting for so much of it to be a hot, drably colored semi-desert plateau, particularly at the higher altitudes.

The tiny vermilion flycatchers which have abounded along the highway today are among the most charming birds I know. Their fiery red caps, breasts, and underparts, in contrast to their black backs and wings, stand out all the more vividly against whatever background they have for the moment chosen. One of their favorite perches is the tip-top-most little twig at the highest peak of a tree. There they shine like living Christmas tree ornaments

Although the Saunderses had a chronic problem with Chico, their new post–World War II modified, four-wheel drive Jeep, it was relatively minor when compared to those inherent in the panel van that George used in earlier Mexican field studies. The latter had far less clearance, was only a rear-wheel drive, and lacked self-sealing tires. (George B. Saunders files)

in the bright sun for hours at a time, leaving every few minutes for a short flight away to snap up some passing insect. Then back, still on the wing, they drop to the exact spot they occupied before, to repeat the process endlessly almost like little wound-up mechanical toys.

I saw my first Brownsville thrasher [Long-billed Thrasher], derby flycatcher [Great Kiskadee], brown jay, and groove-billed ani today. I have seen many colonies of anis along the headwaters of the Amazon River in Peru, but they were in the days when my botanical interests had not had ornithology, and George, added to them. And now, I don't know for certain whether they were the groove-billed or smooth-billed form, or both. Four new birds to be added at once to one's bird life list, as happened today, is a date to be starred indeed.

As we neared Ciudad Mante this afternoon, we saw two little boys, perhaps seven or eight years old, endeavoring to transport a pig from one place to another. They had a small homemade cart consisting of a wooden box on four wheels with a rope at one end to pull it. Into this as we first noticed them one had managed to squeeze Pig, so that his rear was jammed up against the back of the box and his front quarters hung precariously over the front end. Pig was not at all happy and was doing his utmost to remove himself from his predicament. The younger of the two little boys was struggling just as hard to keep him more or less in place, while the other one pulled pretty much both of their weights along the road. Not much progress had been made and it was about a draw, with Pig slightly the favorite to win, as we passed on out of sight.

I have never found Ciudad Mante particularly interesting although it is in the center of a prosperous sugar-growing region. The town itself is very dusty, and it always gives me the impression of being mostly dust-colored too. But there are several comfortable hotels, two of which have swimming pools to help guests withstand the usually considerable heat.

I was forcibly reminded this evening by a walk through town, of something I haven't even thought of during the six months or so since I was last in a Latin American country. The names given to their streets in these lands have always delighted me. It is not particularly unusual to make an appointment with a friend to meet on something called the Street of the Orphans and the Avenue of the Hungry Dogs, or some other name just as odd to us with our less imaginative terminology.

December 24, Tamazunchale, San Luis Potosí

It is Christmas Eve. And here we are for tonight and tomorrow in a little village which is mostly Indio. Its claim among American tourists is its name, which sounds like a fair approximation of "Thomas and Charlie" in English. We had hoped to get farther along the way. But with Chico averaging only twenty to twenty-two mph in country which is not more than gently rolling, we know we shall be in low gear for hours going over the mountains ahead of us. In a passenger car we could have gone on easily. But to need four and a half hours to make each hundred miles changes one's schedule and plans amazingly. We are very anxious to reach Mexico City and a good garage and find out why our speed is so low.

The country today has been far more interesting than that of the past two days. The vegetation has been more tropical from Valles on, and here it is positively lush.

From almost every little yard or patio, the clumpy, tufty branches of mango trees protrude, each a rich red-brown at the end where a new flush of leaves is growing. Or one sees the dusty dull green of an avocado tree against the clattering fringed fronds of bananas blowing in the breeze. Papayas are everywhere, and are among the most tropical looking of all plants in my opinion. The leaves are of a very large size and have deeply lobed surfaces and long, slender petioles. They droop from their hit-or-miss origins along the bare silvery trunks and are extremely exotic and jungley looking. But many papayas as they grow older, or are injured, often put out only a tiny feather-duster tuft of little leaves at the top. These, when coming out above several great green and yellow dangling melons with nothing else below but a spindly naked trunk, give it a ridiculous top-heavy appearance. But however they look, the fruit is delicious, especially when quartered and served with half a lime. The tree, which lives to be eight years old and is the fastest-growing fruit tree in existence, bears fruit when three years old.

I saw a curious sight near Valles this afternoon. It wouldn't have been odd if it had taken place during a spring month, but for early winter it was surprising indeed. Two Couch's kingbirds went through what certainly seemed to be a courtship flight. From their places side by side on an electric light wire they took off together, one slightly below and only a few inches behind the other. With very slow strokes and no effort on the part of the leader to escape the follower, or for the latter to chase away the former, they maneuvered slowly to a considerable height. Then, one apparently clutching the other, or both together, they allowed themselves to fall in a tumbling fashion almost to the ground. Both then perched on the wire again about a foot apart and stayed there for ten minutes or so. One was quiet and the other constantly switched its tail, changed position, and approached the other, then shortly afterward moved away again, making low twittering calls a number of times.

Since the altitude at Valles is only a few hundred feet and the temperature is warm the year around, I suppose it is possible that in some resident birds or other animals, there may not be such sharply marked seasonal variations in sexual activity as there would certainly be in a colder climate.

The bird life in general was rich today. We had fine views of yellow-headed parrots, Alta Mira [Altamira] and Audubon's orioles, the small Mexican crows [Tamaulipas Crows] with calls like twanging Jew's harps, a wood pewee [Eastern Wood-Pewee], phoebe [Eastern Phoebe], an unknown tanager [probably a Summer Tanager], indigo buntings, spotted sandpipers, Mexican goshawks [Northern Goshawks], and Nashville, Tennessee, Wilson's, and black-throated green warblers.

We have both spent Christmases in some odd places, so we won't find Tamazunchale too difficult to adjust to. Its unusual name means "Temple of the Toad God" in the tongue of the Huastecan Indios who inhabited the region before the coming of the Spaniards, and whose descendants still do.

A broad river, the Moctezuma, flows through the little village, which is chiefly a two-street pueblo. There have been built several modern hotels on one side, and a trail with a dozen thatched adobe huts and two tourist courts rim the other. A few years ago thousands of pounds of coffee and other produce left the region at this point on the river by means of very large dugout log canoes. The men sold their cargo and boats, and walked back home to build other boats and repeat the process. But now the road from Laredo to Mexico City goes directly through the town, and the entire picture has changed. Trucks and truck trailers carry what products there are. And the people no longer make or use their dugouts to any great degree.

This evening, we visited the market informally spread along the curb on one side of the street. I found and purchased some delightful figurines fashioned by local artisans that will add to my collection. Further along the street were handwoven wool sarapes for sale. Because of the hot climate, the men do not wear the usual full long sarapes of the high country. They have instead evolved a narrow short one which falls not much farther than their waists, and comes barely to their elbows at the side. We have seen many types of sarapes or ponchos from different countries, and from various parts of Mexico, and each area, and often each town, has its own designs and color combinations which are habitually used. The ones of Tamazunchale are usually woven in dark browns and beiges, and are very attractive. We bargained with one of the native vendors and finally bought a typical one which we can use as a small rug or wall hanging.

At the hotel where we are staying, I have been pretty much interested in the style of machete which the gardener is using. I have never seen one like it before. Instead of the common straight long blade of perhaps two or two and a half feet in length, this type of knife has an overall blade length of less than a foot. And its end is turned abruptly in a curving right angle. We asked the gardener where one could be obtained, and he told us that we would have to buy the steel blade in the local market and then have a woodworker make a handle for it. So while I was choosing my figurines and little animals, George bought a good blade.

Then we started out to find the recommended woodworker's little mud-and-wattle house. This took us to a most definitely non-tourist-frequented part of the town. We wound up and down dirt streets and paths for half an hour or so, frequently asking directions, and finally found his modest home. He was a typical sandal-shod Indio with a pleasant face and amiable manner.

We discussed at length various woods and their grains with him, and eventually chose a fine piece of primavera wood which he had in stock. He promises that it will be all ready tomorrow . . . unheard of haste, and a special favor only because we must leave town the next day. Now we shall *vamos a ver*, as the realistic Latinos say, whether it will be ready or not. I do hope so as it should make us a most useful weed and shrub cutter at home, once it is fitted with a handle so that we can wield it.

December 25, Tamazunchale, San Luis Potosí

It has been a nice Christmas Day. Although after a northern childhood which always included snow and a decorated evergreen tree as an integral part of the festivities, it never seems entirely like Christmas to me unless they are present. We decorated Chico's cab with a Christmas wreath and a sprig of berry-laden mistletoe in the center which we picked in Texas and carefully preserved for today. And we hung our stockings on the bedposts last night.

The traditional oranges and nuts for the stockings were easy to get locally. And I have been hoarding a pound of cellophane-wrapped candies as a surprise for George all the way from Denver, so we had them too. Our major gifts we had already bought and exchanged before leaving home. George's gifts from me were several ornithological books for his library, much too heavy to bring and not needed anyway in Mexico. So they were left behind. But mine from him, a dissecting microscope for which I have yearned for years, can be of great use to us in the field for the identification of bird stomach contents, seeds, and other fascinating things. So it joined the crew under the name of Little Mike and is with us.

For Christmas dinner we chose the most pretentious restaurant in town, a second-story affair where one looks out over the river from one's table.

I had my mouth all ready for roast turkey which one can usually get almost anywhere in Mexico at any season of the year with a *mole* sauce. So we were more than disappointed when not even roast chicken appeared on the menu, and we had to settle for everyday beef.

In the afternoon we took a long walk up the trail along the river's edge. The vegetation is really tropical and interesting, and I saw lots of plants I didn't know but wish I did. There were many gigantic bamboo clumps gracefully bending their feathery tops over the swiftly running water. I never see a luxuriant stand of these or other big grasses without thinking of the botanical taxonomy professor I had at Syracuse University during my undergraduate years. She was a marvel at running down and finding the elusive and wily scientific name of any unknown plant through the intricacies of the keys in Gray's *Manual of Botany*.

But she had a great aversion to the grasses. And she always said that she was saving the study of their identification until after retirement when she would presumably have plenty of time and nothing more pressing to do. So for several years I thought of them as an impossibly difficult group to classify. Then I went on to the University of Michigan for graduate work in botany, and fate decreed that of all things I had to take a course in the taxonomy of the grasses during my first semester there.

I was fully prepared to find the course a horribly unpleasant one. But I was amazed to find it close to fun. It had something of the challenge of a good jigsaw puzzle, or of the tiny pinch of some unknown compound that was presented to each student for identification and the final grade in a course I took in microchemistry. Under the dissecting microscope the flowers of each group of grasses differ from all the others. So their identification is simply a matter of logically following visible clues in a manner of which Sherlock Holmes would have thoroughly approved.

Big black-and-white-headed flycatchers [Great Kiskadees] were abundant along the river today and were voicing their strident *kiskadees* as they darted after insects over the water. Yellow-headed parrots went over in noisy flocks. They always fly by two, like the elephant and the kangaroo of childhood's ditty about the Ark, and with an odd, fluttery wing beat which identifies them as parrots from afar.

We surprised a number of family groups bathing in the river. The children were frolicking like so many wet puppies while the mothers were usually doing laundry or washing their hair. For the women in small Mexican villages one day is pretty much like another, and there is always work which must be done. The campesinos whom we met and talked with along the trail were all pleasant and polite. This is quite a contrast to those of so many tribes in Central and South America who are sullenly or indifferently taciturn.

The men in their clean white cotton pajama-type trousers and loose tops and wide-brimmed straw hats look not only neat, but comfortable. They carry their small belongings in little woven fiber rectangular bags with shoulder straps, much like a child's book-bag. Their loose, unbleached cotton trousers are common to many of the rural people in a large number of countries. They are all tailored according to the same very primitive pattern. Instead of the more conventional buttoned arrangement front and center, they have, probably because of a complete lack of buttons, solved the important need simply and efficiently. Either the right or the left side of the trousers is extended into a large triangle which is crossed over in front so that it reaches to the opposite hip. The trousers are then either tied around the waist by strings, or held in place by a tightly bound wide belt, usually handwoven and of a colorful pattern.

The women are more nondescript in their clothes. But they usually wear full long dark skirts, light blouses, the ever-present shawl which they call a rebozo, and arrange their hair in long braids.

The burros of Mexico make a story of themselves, as they do in every country where they are used. And they are an animal that has my complete admiration and respect. Today I think we have seen specimens of all the varieties there are here. Though I think George really prefers the dark velvety brown ones with buff noses and thick fur. And I find myself partial to the shorter-haired mouse grays which have zebra-like dark brown stripes down backbones and shoulder blades. There are few sections of the country where they are not an important part of the daily lives and economy of the people.

From early dawn to late night one of the constant and typical sounds of Mexico, along with the slip-slap of tortillas being made against bare palms, is the clippity-cloppety tapping of some little donkey's hoofs. Or the nasal and entirely unmusical speech of what some ironic wag has named the "mountain canary."

The little ones are especially appealing. They always seem to be born mostly head and very little body. And of the head, three-quarters seem long twitching ears and cocker-spaniel type eyes. Their extremely soft fur always makes me want to run my hands through it. The few times I have been able to fulfill my wish and pat and hug one to my heart's content, it has been a thoroughly satisfactory experience. Burrito has always stood trustingly still and has even seemed possibly to enjoy it. And Mama has remained courteously polite too. To see a quite recently born baby burro single-mindedly following his mother all day long on the roadside with his little nose usually not more than six inches from her tassel-clipped tail, is an endearing sight.

The adults are used for everything. Panniers are of all types, depending upon the section of the country and the financial state of the owner. They can be wooden, wicker, leather, cloth, or rope. And they are filled with any and everything imaginable from the smaller livestock to tequila plants which are ready for the distillery, to charcoal, or children. Since the peasants usually ride bareback, in order to gain what little comfort they may, they invariably perch precariously on the last possible inch of burro rump. Many a time the odd silhouette of a tiny burro with some long-legged outsized rider stuck in defiance of gravity on the very stern and leaning backward at an acute angle, has made us chuckle as it appeared and disappeared suddenly crossing a road or field in front of us.

Just before dark this evening at the agreed upon hour, we returned to the woodworker's house for our machete. And, *buen hombre* that he is, he had it ready. It is a beautifully worked handle and fits our hand perfectly. I don't suppose we will ever use it in our garden without thinking of Tamazunchale and just how we came to possess it.

December 26, Mexico City, D.F.

We left Tamazunchale early this morning and arrived here in Mexico City just after dark. But we made it in one day! Chico labored up the mountains for hours in low gear, which tiring though it was, did give us an opportunity to see all the passing country, yard by yard, at our leisure. And it is really beautiful.

For a few miles out of Tamazunchale we followed the winding banks of the translucent green Moctezuma River. Scattered small settlements of thatched huts, each one with its clumps of banana and papaya plants and bare-skinned youngsters in the doorways, appeared and disappeared. Then we began to climb, up and up and around and around from 150-foot altitude of the river to 8,600 feet at the highest point on the highway. The road is good, paved, and well banked. And it is not at all hazardous if one keeps on one's own side of the road and toots the horn at curves to remind others to stay on their side.

It was foggy on the hills above Tamazunchale when we left, and we drove through a wet gray belt which opened and shut to allow only tantalizing glimpses of the deep valleys below us. Finally we came out above the fog into the sun and saw one of those glorious sights that usually only those of us who are airplane pilots are fortunate enough to experience. Today in all directions, but especially to the north, rose blue-black peaks and ragged row after row of mountain ranges above a pearl-gray ocean of fog. It was, with its shades of gray and black and the mysterious veiling and half-revealed forms and shapes, like a Doré illustration for Dante's *Inferno.*

For over an hour our views of it changed as we slowly turned and climbed, and it is hard to say which one was the loveliest. Slowly the sun burned away the fog, and finally the depths of the valleys could be seen. At times the road crosses along a narrow, almost knife-like ridge between two mountains with sheer drops of some two thousand feet on either side. And in some of them, blustery though it is as the wind whistles through the natural gap, the rural folk have chosen just that spot to build a little village.

At each of them the one row of small houses looks as though glue must have been needed to stick them in place. Any wandering pig or chicken which takes a single misstep must have a long fall indeed before being retrieved. I can imagine that they have to tether the very small children by long ropes to a house pole, as the barge and houseboat dwellers do in the United States and other countries.

The high pine- and oak-covered mountains were refreshingly cool after the humid heat of Valles and Tamazunchale. Bushy-tailed squirrels scampered from the rocks and before us across the road. And brilliant jays [probably Mexican Jays] with bright blue backs and grayish underparts hopped below us in the treetops. Finally we were through the really steep wooded mountains, and the Cañón de los Mármoles, "Canyon of the Marbles," which is the second-highest point on the entire Laredo–Mexico City highway. Abruptly we came into a hot, cactusy, arid belt which is still high and hilly.

Here we entered the territory of the Otomí Indians, one of the poorest tribes in Mexico. One of their principal plants is the giant agave known as maguey. From it they obtain two important things: an alcoholic drink, pulque, with which large numbers in any village try daily to forget their troubles; and a tough leaf fiber from which they make thread and rope and weave useful articles.

Their lives are difficult, and difficult in the extreme. The ever-present pulque that provides liquid and a certain amount of nutrition too, eventually causes illness and death. But while they live, they also use its leaves as wall and roofing material for their tiny one-room huts. Many are so small that one must stoop to enter. There are usually no windows and no doors that close. At night the bitterly cold mountain air is free to enter through every chink and interstice. And it must make sleep nearly impossible, for many of them are too poor to own such a thing as a sarape or a blanket. In fact, it seems a miracle that the human race, with so little to help it out, can survive there at all and that the very high mortality rate isn't higher than it is.

They are a small, short people, and both men and women are commonly dressed in white homespun garments. Small herds of sheep and goats seem to make up their only agricultural life. And with the extremely limited water supply where they live, they do well to keep them, or even to have enough for themselves. Dozens of men and women, from the eight-year-old goatherders to the most wrinkled and tattered old crones, carry distaffs and make thread from a fluff of agave fiber as they walk along the roads. Some stretch the yard-long hanks over one shoulder. Others drape them on their heads so that when come upon unexpectedly around a curve, they look as though masses of light blond hair, tastefully arranged in high pompadours, grace their otherwise exceedingly drab and ragged costumes.

Cacti of every shape and form grow everywhere, blooming in cerises, reds, oranges, and yellows. And the Otomís make use of them too. The stems of the columnar types are cut and stood upright side by side in the dry soil to form fences or the walls of houses and small corrals. The fruits of the nopal cactus are gathered and eaten. And the young green "leaf" pads of the *Opuntia* are assiduously searched for and brought home by the small children to provide a meager supper for a family. Two or three flat pieces are usually stuck on a little twig to avoid the sharp thorns.

Where spiny cacti or euphorbias aren't, thorny leguminous shrubs and trees are. A botanist who specializes in desert plants would have a glitter in his eye, indeed, and a wonderful time here digging and hacking and being scratched by them. But hot, arid country and its vegetation did not now bring forth this response in either George or me, so we moved on at our best speed.

Two small unexpectedly green valleys in the area with fine irrigated fields of alfalfa and wheat contrast strongly to the poor, barren, beige-colored country around and between them. They are a forceful example of what water and industry can do with even this land, providing one has some capital to start with.

We reached the town of Zimapán, which is interesting for being the spot where the Spaniards after the Conquest opened their first silver mines in 1522, and for the very old church on its central plaza. Farther along the way, to the left of the highway, is the Convent of Actopan, which dates from the sixteenth century and which has fine lines outside and some magnificent frescoes inside. The landmark of the Cerro de los Frailes, "Hill of the Friars," appeared soon afterward. This series of highly eroded rocky spires on a nearby mountain top looks from the distance of the road as though it might indeed be a group of gigantic friars preaching with uplifted arms.

We chugged up to what our altimeter registered as the highest point on the road at eighty-six hundred feet, and then dropped down slightly to enter the end of the old lake bed which at one time filled much of the Valley of Mexico.[5]

For forty-five miles we drove along its dusty, windblown length with enormous dancing "sand devils" whirling their diverse ways around us. The last twenty-five miles or so are extremely interesting, as the old stone pillars of Spanish colonial days, each marked in the number of leagues it is distant from Zócalo, or the central plaza in Mexico City, still stand at one side of the present road.

The modern road parallels the ancient one, and the colonial stone bridges with their small carved statues and name plaques today stand only a few feet away in many places. Just outside of Mexico City the present road still follows the top of the narrow causeway which the Spanish conquerors built across Lake Texcoco when it was water filled.[6] And the high stone wall which held back the waves to the windward of the road is now just a yard or two to the right of one's elbow as one drives into the city. It makes the twentieth century seem not so far away from the sixteenth century after all.

December 27, Mexico City, D.F.

We are settled in a very pleasant small hotel on the Calle de Lerma just off the Paseo de la Reforma, which has a gorgeous flower-filled garden and lots of birds. George is seeing the proper authorities and unloading things from

Chico that we will not need and repacking for our first long field trip.[7] And I intend to bask in the sun in the garden and catch up on my letter writing since I owe letters to almost all of my friends.

Among the numerous birds in the garden which I am enjoying watching are several song sparrows. They are handsome fellows and are marked much more darkly than any others I have ever seen before. They also have the added oddity of constantly holding their tails cocked up at an acute angle like so many oversized house wrens. They are a busy lot and endlessly scurry about under the roses and begonias scratching up the soil like a flock of Lilliputian barnyard fowl. They scratch so rapidly in fact that to the naked eye they appear to be doing it with both feet, simultaneously, no mean feat.[8]

Many evenings when we have no birds or plants to prepare and when we are in a large enough place to have light, it is our pleasure to have a bout of rummy.

December 30, Mexico City, D.F.

We took a day's field trip today out to what is left of Lake Texcoco, to estimate its duck population and to see what kinds of waterfowl there are on it this year.[9] Although in the days when Cortés and his Spanish conquistadores first saw the spot that is now Mexico City, it was on one of a series of islands in the center of a beautiful lake, all that has long been changed.

The lake was almost at once made smaller by the digging of a series of large deep canals for the express purpose of draining it.[10] Through the centuries this has continued.[11] More and more of its water has been diverted for agricultural irrigation and for the use of the city. And the nearby hills have been almost entirely cut over so that their trees no longer protect the watershed and add water to the lake. Mexico City today lies in a large arid, windswept bowl, and the air is yellow much of the time with the irritating blowing dust of what was once the water-filled lake bed.

A tiny remnant of the lake, all that remains, lies east of the city, and each year even this becomes visibly smaller.[12] Much of it is divided into small units by dikes, and a part of it is controlled by a chemical plant. We drove today out Route 1 to the Venta del Carpio and turned off on the side road which leads to the justly famous pyramids of San Juan Teotihuacán. At one point on this road a part of the lake comes within sight, so we parked Chico off the road and started to walk over the arid, uncultivated ground to the lake.

We could see that ducks were there, but we were still too far away to identify them when our way was suddenly barred by two men who brusquely announced that we could go no farther and that we were on land owned by their *ejido*. In the early 1930s when the large private estates or haciendas were taken away from their owners by the Mexican government, they were almost all turned over as communal properties, or *ejidos*, to the native farmers who had formerly worked on them. They are now worked cooperatively, each man theoretically working equally and receiving an equal share of the proceeds of the harvest.

We explained who we were, what we were doing, and that we had papers from the Mexican government authorizing our work anywhere in the country. We also told them that we had no wish to do any hunting; we simply wanted to go nearer to see what kinds of waterfowl there were in that end of the lake.

Nothing did any good whatsoever. They were soon joined by several others from a reed hut half a mile away, and it was evident that there was nothing else to do but to make as dignified a retreat as possible under the circumstances. Neither of us saw any point in attempting to continue by force. But it had been a rude surprise, the more so because it was entirely unexpected. If we find that attitude in very many of the places where we plan to study waterfowl, it will make a tremendous difference in what we can accomplish.

So we are not very happy, although we don't see how we could possibly have done anything to change events. The men didn't want money from us. They simply, for some reason, wanted no strangers around their land or lakefront.[13] It is the first time in all the years that George has worked down here that such a thing has happened to him.

When we were thoroughly convinced today that we were not to be allowed near the lake and that we could do no fieldwork after all on that end of Texcoco, we drove on the short distance to the archaeological ruins of San Juan Teotihuacán, since we were almost there. Teotihuacán means in Nahuatl, "Place Where the Gods Live," and it was in former times one of the most sacred places of the Nahoa religion.

As in so many Mexican ruins there have been discovered still older buildings beneath those of the Aztecs. It is thought that these were built by the Toltecs, but at just what date no one knows with absolute certainty. There are two great main Pyramids of the Sun and of the Moon, which rise high above the ruins of a city that at one time spread out from their bases some four by two miles. They are considered to be the two largest artificial mounds in America and were made of adobe blocks faced with stone.

The Pyramid of the Sun is my favorite building of all the ruins. From its base there is a magnificent view straight up the great reconstructed light-colored stone face, to a brilliant blue sky above. It can't help but put one into the proper perspective of one's minuteness on the face of the earth. It is the usual truncated pyramid shape and measures about 215 feet high, and 720 by 760 feet at the base. At the top it narrows to a 105- by 60-foot flat summit.

A straight flight of 222 steep steps, but nothing so difficult as the uneven blocks of the Great Pyramid of Cheops in Egypt, leads from the base to its top. We climbed up them for the view, and because the top seemed an ideal vantage point from which to take photographs. With sharply aching leg muscles we made it, and the sight was well worth the effort. To the west stretches the paved walks of the Highway of the Dead, from whose many tombs hundreds of small clay heads or masks, supposedly effigies of buried priests, have been discovered.

About half a mile to the north stands the smaller Pyramid of the Moon, also terraced like that of the Sun. It rises about 150 feet, and is some 425 by 510 feet at the base. There is a small flat summit approximately 20 feet square. Both of the pyramids have been built of five separate layers of earth and volcanic rock, each layer complete in itself. The outer one was originally plastered over with a coating of fine cement or mortar. And if it was anything like the Maya temples of Chichén Itzá in Yucatán, there were probably sprightly paintings and frescoes in bright reds and other colors at one time on this outer surface. I couldn't find out today from anyone there whether such drawings have ever been found on the pyramids.

A mile to the east is a very interesting ruin known both as the Temple of Quetzalcoatl and the Citadel. It is a great rectangular court with minor pyramids within it. But the most interesting thing about it to me is the very fine stone carvings of mythological subjects which have been discovered on one wall.

The usual small Indian urchins were stationed at strategic points to sell, if possible, small terra-cotta figurines and heads which they swore were *legítimo,* and from the ruins in fact. Although probably 99 percent of all the objects sold to tourists in Mexico as *legítimo* archaeological remains are patently a few months old at the most, these fragments really looked authentic. And some of them may well have been.

To me, one of the most worthwhile things to see at the ruin is the Regional Museum, which the government has established there. In it are excellent relief diagrams of all the buildings with the approximate age and culture that built each one. And there is a splendid general history of the entire region. We spent most of our time hanging over the many cases of fascinating little masks, jadeite and obsidian articles, and jewelry which have been excavated from the sites, all neatly and accurately labeled.

Finally we started back toward Mexico City. Just outside the Zona Arqueológica we stopped Chico so that I could take a panorama in color of the entire Pyramid of the Sun framed by *Opuntia* cacti in the foreground. I climbed up a steep road cut to a slight rise of land for my picture, and George idly started to examine the earth and stones in the road cut. This interest in stones is a permanent hangover from his late high school and early college days. Then he first developed a great interest in mineralogy and spent summer vacations roaming the Rocky Mountains in an ancient Model T Ford, geology pick in one hand.

A call from him soon brought me back. He held triumphantly a small black obsidian knife which he had just dug out of the loose earth. We both searched opposite sides of the road for half an hour or so, and found a total of about a dozen of the knives, fragments of a highly decorated pottery, a large terra-cotta ring made for a purpose I can't imagine, and a terra-cotta bead or two.

In terms of value they haven't any. But we had a wonderful time searching for them and thinking that at any moment some exciting unusual object might turn up. Although I suppose to be really honest, I must admit that we found these bits fully as exciting as anything could have been. And at least we know they are authentic and weren't fashioned a month or so ago by some agile but not too honest brain and hands.

December 31, Mexico City, D.F.

Today we decided to try different tactics in studying the ducks on Lake Texcoco. We took along with us a member of the Mexican federal wildlife service. He is a middle-aged man who for a great many years has worked in the Distrito Federal and knows and is known by a great many of the farmers and fishermen outside of Mexico City. It worked like a charm. Sr. G. knew exactly the place where a native family lives on another edge of the lake and where, through his introduction, we are welcome to observe the waterfowl to our heart's content.[14]

The family lives in a small one-room reed hut with a thatched roof, alongside a narrow canal leading inland a hundred yards or so from the main body of the lake. At the time we visited them, several tiny handmade dugout canoes were tied to a tree, the smallest being a one-man affair some four and a half feet in length, ten inches in width, and eight inches or so in height. We stepped gingerly down into and up out of this small one from the three-foot higher walls of the ditch, using it as an extremely jiggly bridge.

The father of the family sent his twelve-year-old boy along with us as guide, and we walked half a mile or so along a raised foot-wide earthen dike through mosquitoey marshes until we reached a small quiet bay. The ducks were there all right. In the single small area we saw more than two thousand gadwalls, baldpates [American Wigeons], shovelers [Northern Shovelers], pintails [Northern Pintails], and cinnamon and blue-winged teal. Some preened feathers or floated seemingly in a doze.[15] Others up-ended, leaving only triangular tail patches above the water as they lunched on succulent aquatic plants just beneath the surface of the shallow bay. Occasionally several were moved to splatter off to join friends elsewhere. And their places were taken almost immediately by new arrivals that landed with a series of mighty splashes as they skidded to a halt.

After lunch we continued away from the city for another twenty-five miles to Lake Zumpango to see what waterfowl might be there. The lake area is very large, but the water is extremely shallow, and each year it is becoming more so. We stopped at several places along the shore and looked for ducks or other waterbirds through our binoculars, but not a single one did we see.

A few white pajama-suited men with trouser legs rolled above their brown knees were wading in the shallow water and fishing. Their nets are held out rigidly in a semicircle by bent poles, and a long handle is attached to the middle. The men walk slowly along pushing the nets before them, much as though they were using suburban lawn mowers. Such small fish as are caught are later transferred to wicker baskets onshore.

We talked with one fisherman, and he told us that ducks are seldom seen on the lake now, although in former years there used to be many. The water is now so low that it is difficult to catch enough fish to feed a family, because so much of it is being drawn off through the large canals to supply Mexico City. Before many more years go by, this whole area will probably be just another one of the big dry, flat plains of which Mexico has altogether too many now, instead of its former beautiful useful lake. And the living standards and diet of its nearby inhabitants will be reduced still further.

CHAPTER 2

To Lakes Cuitzeo and Pátzcuaro

January 3, 1949, The Camp of the Black Robin, west of Toluca, Mexico

WE LEFT Mexico City about two o'clock this afternoon and are off on our first real work in the field. George is in khakis and I in khaki shirt and skirt, and since this is Chico and Chiquita's color scheme too, we all blend in nicely and look, we think, quite official. We are quite saddened to find that Chico's speed has not improved a bit in spite of his innards having been prodded here and pushed there and an expert opinion given that the governor on his engine is not functioning in any way.

We crossed the mountain pass west of Mexico City this afternoon at a little over ten thousand feet in altitude before dropping down into the valley of the Lerma River. Hail was falling and a light dusting of snow covered the more deeply shaded sides of the road. The view was a beautiful one. The pure white snow-covered mountain peak called the Nevado de Toluca rose on our left to over fifteen thousand feet, and the reddish-brown plain stretched out beneath us with a few towns scattered here and there and the city of Toluca directly ahead. After inching our way through the narrow, congested streets of Toluca, we were happy to be on the open highway again. We continued along over rather barren arid country until nightfall caught us in an area of high, wooded hills. After a few minutes' search we fortunately found a place where we could drive off the road, and so we are camped in a lovely oak-pine forest at a little over eight thousand feet in altitude.

Now that darkness has fallen, it is very cold. Our little nylon tent is set up, and a table in front of it has food, a small two-burner gasoline stove, and a lantern on it. We had decided not to use the lantern in an effort to avoid attracting local company, since we were near a main highway. But campesinos appeared five minutes after we arrived and have remained since, so it couldn't draw more.

Our visitors are all men and boys, wrapped in heavy wool sarapes that cover their bodies and faces up to the nose line. They look like so many mummies as they silently stand or squat in the shadows and with fathomless black eyes follow our every move. They seem neither friendly nor unfriendly, but more like such stolid elemental forces as the rocks and trees around them.

As we were setting up camp just before nightfall, I saw my first black robin flitting through the pines. They are about the size and shape of our familiar robin redbreast, but are entirely black. We have decided to name each camp we make with the name of an outstanding bird seen while we are there, so this will be our first one.

Tonight the stars are very nearly the most brilliant I think I have ever seen. And in spite of our uninvited guests, we haven't been able to resist sitting for an hour or so and just enjoying the sky before turning in. We are both fairly at home with the constellations of our familiar New York–Denver latitudes in the States, George more than I. But they are of course very different here. And our attempts to identify some familiar celestial landmarks this evening weren't especially successful. I am afraid we will have to start and learn them all over again . . . if we can find any sky charts for this part of the world.

January 4, The Camp of the Painted Redstart, near Morelia, Michoacán

The sky was an electric blue this morning with long thin streaks of purest white clouds, so high I knew they were of the cirrus variety and really made up of ice crystals. The deep intense blue of the skies of the highlands of Mexico and Central America is to my mind far lovelier than the softer and more often praised paler blue of Italy and southern France. To me this deeper tone is more powerful and stimulating.

After breaking camp, we chugged over mountains and valleys all day at

The Saunderses' field investigations in remote locales frequently required them to travel very poor roads or trails. This photo taken by George on an earlier trip demonstrates the hazard of crossing a steep-banked arroyo. The winch, mounted on the front bumper, was useless without a tree or large rock to which the winch could be attached. (George B. Saunders files)

our usual pace. Late afternoon found us in a pleasant little spot many miles short of the city of Morelia. Chico's four-wheel drive is wonderful for driving off and down into desirable camping spots out of sight of the road. We are now under several big trees on a grassy slope, and along the farther side of the road a free, wild little stream bubbles along.

After the tent was up and while I was operating the two-burner stove and juggling three hot kettles between them, George strolled off to see what birds he could find. A shout to come running at once made me leave kettles and food, temporarily, to their fate. However, by the time I arrived, the bird had flown. It was a painted redstart, which I had never seen, and still haven't. But tonight's camp has already named itself.

The patch of woodland, however, shortly afterward redeemed itself in my eyes because while preparing dinner, I saw in it several Audubon's, the black-throated gray, the orange-crowned, and the olive warblers, and a black phoebe, all for the first time. And we have had here this evening one of the strangest and most enchanting experiences either one of us has ever had.

Since the night was very dark and we seemed far away from any inhabitants, we decided to use the Coleman lantern. George had an hour or so of work he wanted to do, and I had plant notes that very much needed to be brought up to date. So we lighted the lantern, which cast its usual brilliant white light over us and the camp.

We were still standing with it just lighted in George's hand and I beside him, when suddenly a tiny bird fluttered directly to it out of the darkness. Deftly it snapped up a small insect and alighted on our table edge to swallow it and wiped its bill. Then it repeated the maneuver over and over again with a series of vertical banks and zooms that any airplane pilot would have envied. As it spread its wings and tail fanwise almost constantly with little flits, we knew from the beautiful yellow patches against the darker plumage that it was a female American redstart.

For ten full minutes by my wristwatch the little beauty continued to dart around our lantern feeding. She paid not the slightest attention to us although we must have been perfectly visible in the strong light. And all the time she put on her fascinating display for us like an oversized butterfly; her bright black eyes sparkled in the light as she twisted her head from side to side in deciding which bug to next dash after.

The highlight of the whole affair was to see her after one flight land confidently on George's hand as he still held the lantern, and for a few moments afterward to have her actually alight on my shoulder. She stayed there for half a minute or so, a nearly weightless mite. I would hardly have known she was there at all if my head hadn't been turned in her direction so that I was looking at her from only a few inches away. Utterly fearless and trusting she was. And her touch made a chill go up my spine and my back hair rise, the way it does when I hear supremely bewitching music. It was all eerie and wonderful and almost other-worldish, the setting contributing with its tremendous contrasts of shadow and light.

Finally after a very substantial moth and beetle supper she flew over to our table again. She perched on its edge for a moment or two while she cleaned her bill thoroughly and fluffed and preened her feathers. And then off she flew into the outer darkness from wherever she had come, and we saw her no more. But I doubt if either one of us will ever forget her or the tingling thrill she gave us.

January 5, Morelia, Michoacán

We almost froze last night, and the temperature on Chico's thermometer *inside* the cab was under 40°F at eight A.M. today. Our little valley illustrated perfectly a principal of meteorology that both George and I had completely forgotten, that cold air runs down the slopes of hills and settles in pockets on the lowest land. And the lowest land was just where we were camped. It took two dishes of piping hot oatmeal and milk, and three cups of coffee apiece, to even start to thaw us out this morning.

When we were just beginning to be comfortable and hot dishwater was bringing back some feeling into my numbed fingers, the air was suddenly filled with a tumbling cascade of tinkling notes. It had a ventriloquistic quality as though all around us at once icicles were dropping and sliding down a frozen icy slope.

Over and over again the melodic voice lifted in its rippling song, and we knew that a [Townsend's] solitaire was greeting the morning. It is the loveliest bird song I know, bar none. I am only sorry that this thrush is such a shy bird and found usually in rugged terrain at high altitudes away from habitations, so that few people hear it as a daily part of their lives. It is true that they are frequently taken from their nests when young and sold as caged birds in the cities, where their song is often heard.

We drove on the few miles into Morelia shortly after daylight, and then turned north on a road that was shown as paved on our map. It turned out to be the dustiest, roughest sort of washboard, and honeycombed with deep holes. To save the tires and remain more or less in our seats, we took the twenty-five-odd miles to Lake Cuitzéo at the pace of an agile snail.

It is some of the most depressing country I have yet seen in Mexico, arid and barren with practically no agricultural crops nor surface water. How the inhabitants of the occasional villages scrape together enough water to live, we could only guess. It must be by wells, although we looked into many yards and streets as we passed and saw not one. In the rainy season they doubtless grow corn and pumpkins in enough quantity to store them for use the rest of the year, but they can't store water that way.

Lake Cuitzéo is about thirty-five miles long and some three miles wide at the place where the road crosses it on a stone causeway.[1] Here it is only an inch or two deep in most places, and dozens of snowy and American [Great] egrets were sedately posing over its surface, seeking brunch in the muddy waters. No ducks of any sort were to be seen nearby, but many were rafted farther to the west.

Several fishermen, with white trouser legs rolled up dark legs to their muscular thighs, walked about here and there in the lake, mostly along slightly deeper channels near the causeway. But all of them were not more than ankle deep in the water. They pushed their long-handled curved nets before them to scoop up the fish, but none seemed to be catching much.

A group of chatting women washed clothes at the far side of the lake. With native women the beach of a lake, or edge of a stream instead of the plaza of a town, is often the hub of their social life. There they wash, each on her knees or standing knee deep in the water, bent over her favorite flat stone. On this she scrubs or rolls or pounds her family's washing, gossiping with her other neighbors in the row as she does so.

The people of the town of Cuitzéo, which is now left high and dry half a mile away from its shore by the receding waters, are still fortunate to have any water at all near them. When the western part of the lake eventually dries up, as it doubtless will within a few months, their life will be even more difficult than it is now. We had planned to put Chiquita in the lake here, attach our five-horsepower outboard motor at the rear, and cruise over the greater part of its surface to do a thorough waterfowl survey. But the water is too shallow even for her slight draft, let alone the propeller of the motor. So we had to retrace our steps to the first main road going east shortly before arriving again at Morelia.

We turned off on its gravel surface and drove for more than an hour on it to a point where our map showed that the road comes to within a mile of the easternmost end of the lake. A number of futile tries to get near it by turning off on several dirt roads got us nowhere. Each one ended either in impenetrable marsh or merely on the edge of now dry bays still far from the water. Eventually we finally managed to find one spot on a bluff off the road where we could see the eastern end of the lake. There we saw some baldpates, gadwalls, shovelers, canvasbacks, [Lesser] scaup, and blue-winged and cinnamon teal. So there are ducks still on part of the lake, although it is now impossible to do any sort of a satisfactory inventory of them except by air.

We soon retraced our path to the main Morelia-Cuitzéo road and along it to the town of Cuitzéo again. Then we continued still northward for another seventeen miles to Lake Yuriria, through country still more arid if this could be possible. This lake is a deeper one and on it George estimated there were a total of about forty-five hundred ducks, mostly baldpates, shovelers, gadwalls, Mexican ducks, and blue-winged and cinnamon teal.[2] We came back through the town of Yuriria, which is more formally named Yuririapúndaro, meaning "Lake of Blood," though it is now never called by the longer name.

It is a large and interesting place and dates from long before the Conquest. It still has several fine old Spanish colonial churches, and beautiful handwoven sarapes hang for sale in the doorways of some of the tiny shops. We had no time to stop and bargain for anything, as I would have loved to do, since we had to reach a hotel in Morelia by nightfall. So we pushed Chico to his best pace and arrived here just after dark.

Morelia, at sixty-two hundred feet in altitude, is an old city. A fine old aqueduct, built in the 1780s by the Spaniards, still brings water into the city from distant mountain springs, and startles a visitor on his first drive from Mexico City as it suddenly appears and divides the main road in two.

January 6, Pátzcuaro, Michoacán

The thirty-mile drive to Pátzcuaro is most scenic, especially after the deep blue lake and its towering mountains come into view.[3] Near the eastern edge of Lake Pátzcuaro is the village of Quiroga, famous for its colorful displays of lacquerware and tiny, woven-seated wooden chairs, all decorated in bright colors. I couldn't resist one in the market of bright yellow, trimmed with scarlet flowers to go by our fireplace at home.

From there our road bordered the southern shore of the lake. At frequent intervals we stopped, George to study the waterfowl while I watched the colorful, friendly Indians who passed.

One of our stops along the way was the outskirts of Tzintzuntzán, a town midway between Quiroga and Pátzcuaro, to negotiate the purchase for our collection of curia of one of the small lollypop paddles used locally by the Indios on the lake. A ten-year-old boy was standing along the roadside with one which had weathered to a beautiful mahogany color, though they are carved in one piece from pine wood.

We asked if he would like to sell it and what the price would be. After a brief conference with his mother in a nearby adobe one-room house, he shyly and politely said that they would sell and named a reasonable figure. So we bought it and left the small salesman a piece of candy as his commission in the deal.

The land all around this area is cultivated or occupied by small dwellings, so there are no suitable camping spots. However, a few years ago a very comfortable posada, or inn, was built just outside the village of Pátzcuaro and nearer the lake. It is constructed in the traditional Spanish colonial style around two inside patios and overlooks the lake from its upper story.

The front patio boasts a fountain, and its more expensive rooms are usually taken by tourists on prepaid tours. The back patio is fountainless, but also flower-filled, and is considerably lower in price. Also it is quieter. So we are well established in a room with bath on the back patio, and plan to spend several days here while George studies and estimates the waterfowl on the lake and I delve into the mysteries of their plant foods.

The posada is set in rather extensive gardens, which are planted with dozens of rosebushes and a number of fruit trees of various sorts. They make a fine gathering place for the birds of the neighborhood, and we were treated via our back window to a very pleasant concert of avian vocalists late this afternoon. House finches, or linnets, easily led in the affair. Their bright rose-red breasts, foreheads, and rumps gleamed brilliantly in the setting sunlight as though reflecting the color from the coral-pink cirrus clouds high above them.

January 7, Pátzcuaro, Michoacán

Chiquita has been duly baptized in the Lake of Pátzcuaro today, and she behaved perfectly. And what is just as important, the outboard motor behaved perfectly too. With much heaving and groaning we lifted the canoe down from the top of Chico and carried it a hundred yards to a canal which enters the lake and is closer to the road. Then while I carried binoculars, leather jackets, plant vasculum, guns, and hunting bags draped here and there around my person to the intense interest of the local campesino onlookers of all ages, George went back to the truck and got the motor and fuel can.

Everything functioned correctly at once, and we were soon skimming over the surface of the lake at a good pace. George estimated it to be at eight knots with full throttle, and sitting as we were close to the water, it seemed every knot of it. As we overtook and passed the clumsier rowboats and motor launches that commercially go back and forth from Pátzcuaro to the island of Janitzio in the lake, we collected a fine assortment of surprised and envious looks.

The natives who paddle into the village in cumbersome dugout log canoes with paddles the shape of and about as efficient as round lollypops, and from their little towns elsewhere around the lake, simply stared in unbelief. Chiquita may well have been the first factory-made canoe seen there, and doubtless was the first all-metal one with trim efficient lines and speed. Another cause for their staring was to see a boat moving along with the man in it apparently arranging the motive power and a mere woman simply sitting at ease enjoying herself. In most of the dugouts we have seen, the women have worked steadily at the paddles while the men sit grandly at one raised end, literally above such menial tasks.

The little fishing village of Janitzio on the island of the same name is a most interesting place. As we cruised along its shore, rows and rows of light brown handwoven fishing nets were stretched up its rocky slopes. Children played everywhere in dugout log canoes, which were drawn up on the narrow beach. And brightly skirted and bloused women knelt at the water's edge to scrub their ubiquitous laundry on the rocks. Above the red-tiled roofs and the spire of the old church looms from every angle the white cement figure, in gigantic scale, of the patriot Morelos, holding aloft one mighty arm in silent benediction over the scene.

We had been told that a strong wind usually arises daily about ten A.M. making the lake unsafe for small boats. But today we kept merrily on our way until mid-afternoon and had no difficulty whatsoever. I saw my first pied-billed grebe, and we collected one to make into a scientific study skin for the museum.

Grebes are among the most primitive of the bird groups, and their feathers feel like thick fur when stroked, and even look like it when examined closely. Their bills are exactly like flexible rubber. Here they are wary birds and have an irritating habit of diving and swimming underwater for long distances before one can get anywhere near them. So only a well-placed, long-range shot secured this specimen for us.

We also saw considerable numbers of coots [American Coots], shovelers, baldpates, gadwalls, blue-winged and cinnamon teal, ruddy, scaup, and canvasback ducks, and American bitterns on the lake. And along the shore were [Loggerhead] shrikes, Cassin's kingbirds, lark sparrows, marsh [Common] yellowthroats, and Brewer's blackbirds.

We returned to the posada well pleased with our birds. But few water plants are turning out to be in flower at the present season and so aren't worth

A Tarascan Indian in a dugout on Lake Pátzcuaro, Michoacán, on July 14, 1904, photographed by Major Edward A. Goldman. Goldman; his brother, Luther. J.; and son, Luther C., sometimes known as the "Goldman Triumvirate," were early federal ornithologists who conducted natural history studies in Mexico several decades prior to the Saunderses. Dorothy noted that dugout canoes powered by "lollypop"-shaped paddles were still used on Pátzcuaro in 1949. Many rural Mexicans were wearing the same white cotton clothing in 1948–49. (Photo by Edward A. Goldman, U.S. Fish and Wildlife Service, from George B. Saunders files)

collecting, as they can't be identified. We had the back patio to ourselves, so we moved tables out into it and worked in the sun until dusk. Because of the high mountain range to the west of the lake, darkness comes suddenly. And with the setting of the sun the temperature drops abruptly at this altitude of 7,180 feet. Within a half hour's time I shifted from a short-sleeved cotton dress to a sweater and wool coat, and I was still cold.

But there are always big fires in the evening in both the dining room and living room fireplaces, so it is very comfortable here. The lake is noted for a small white translucent fish, locally called *pescado blanco*. It is a peculiar fish with flesh so colorless that one can actually see the main spine and larger organs through the skin. The largest I have seen reached no more than a foot in length. The posada prepares the fish for meals by rolling them in a mixture of flour and beaten eggs and fried in deep fat. This delectable dish is served twice daily with a picante sauce, and we are very fond of it.

January 9, Pátzcuaro, Michoacán

We have decided to rename the posada the "Camp of the Canyon Wren," even though strictly speaking it isn't a camp at all. Each morning when the first rays of light reach the second story of the back patio, a canyon wren who sleeps on one of the pillar tops and who is now known to us as Carlos, starts his liquid silvery run down the scale. It reverberates around the four walls below him as though a dozen birds were singing their hearts out. Five or six times he sings the melody, head up and white throat swelled and vibrating, and then is off for the day not to repeat it within our hearing until he returns at dusk to bed.

It is one of the loveliest bird songs I know, and ranks in my estimation a close second to that of the solitaire. The local name for him is *saltaparedes*, meaning "He Jumps Walls," which doubtless refers to his activities during the day when catching insects for food around farms and homes. Wrens seem to be common in this neighborhood. Besides the canyon, we have seen several rock wrens darting with flicking tails in and out of the piled stone walls along the roads of the nearby villages, a number of tiny house wrens, and two unknown (to us) little fellows in hillside shrubs.

We spent yesterday again out on the lake in the canoe. And today we took Chico and drove as far as possible around the western side of the lake to estimate the duck population in the several bays there.[4] We were told when we inquired that the road is "muy bueno, puro plano," fine, very flat. But we certainly wouldn't have liked to take a passenger car on what we found. On one steep cobbled hill we had to use the four-wheel drive to make it. However, it was a most interesting trip, and we saw a good third of the shoreline of the lake and the waterfowl there.

The towns we went through are old, and several have lovely ancient Spanish colonial churches still being used in them. And we are invariably finding the native inhabitants here friendly and helpful. Having lived or traveled in all but two of the twenty-one Latin American republics, and having seen many examples of their various indigenous tribes, the Tarascans who live around Pátzcuaro are so far as fine as any group I have encountered. So many I have seen in the highlands of Peru, Ecuador, or Guatemala, especially, have been stolid indifferent people, unsmiling and unresponsive when spoken to, bowed down by the problems of their hard lives.

The Tarascans have fine features and a light brown skin, and are the only natives I have ever seen who spontaneously whistle tunes when walking along the roads. We talked with a number of the pure-blooded ones we met today along the lake . . . an old woman waiting with a little girl for a bus . . . a boy with a burro . . . a farmer, and others. They are intelligent and their information was accurate. And surprising to me, they carried on a conversation as such, instead of merely answering a question and lapsing into silence immediately thereafter.

Most of the women still wear the old-style costume of handwoven wool skirt, either bright red or dark blue in color, a simple blouse, and the alternatively striped dark and light blue rebozo or shawl of the region. The men have unfortunately taken to store pants and shirts, but still wear their handwoven wool sarapes except in the heat of the day.

I have never seen such brightly colored vermilion flycatchers as fairly bead the wires and fences here. Straight up into the air as though it were a helicopter a little one will rise, snap up a passing insect, and drop light as down on exactly the same twig or spot he left. Over and over again the maneuver is repeated in a glorious display of crimson and black. I saw my first yellow-headed blackbirds today when two flocks of them flew along the road close to us. They are beautiful shiny birds. But I think I like the great-tailed grackles better, although I know what scoundrels they can be. They are larger and glossier and have more personality.

CHAPTER 3

Chapala and Guadalajara

January 10, Camp of the Black-necked Stilt, Lake Chapala, Michoacán and Jalisco

WE DROVE to Lake Chapala today through what is in my opinion some of the most interesting country in Mexico. How often I wish for more than the one year of geology I studied in college. Pátzcuaro and the region west of it were areas of extensive volcanic action in ancient times. And there is now one new volcano, Paricutín, which originated in February 1943 in the small cornfield of an astonished campesino named Dionisio Pulido.

Its base is seven thousand feet above sea level, and the cone now rises some three thousand feet more. It is growing steadily through its irregular eruptions. The nearby towns of Paricutín and San Juan Parangaricutiro have been completely covered with the hardened lava. Of the latter, only the old church steeple and dome rise eerily above the flow to mark the former location of an active village. The windblown volcanic ashes have drifted for miles over the countryside, making it at present completely unfit for agricultural use.

The trip in from Uruapan to the volcano, the last forty-five minutes of it by peasant-led horses, is a wonderfully interesting experience which I had

In 1943, a Mexican peasant was astonished to see a lava flow emerge from his cornfield, a few miles northwest of Uruapa, Michoacán. He was witnessing the birth of Paricutín, a new volcano. The extraordinary event drew worldwide publicity. In its first year, the volcano rose to about 1,390 feet. The volcano continued to heighten until 1952 when it had attained elevation of some 8,995 feet above sea level. Here Paricutín is shown in a state of minor eruption. Miraculously, no human deaths resulted from the new seismic activity. January 1948. (Source unknown)

during a former stay in Mexico. Late afternoon is the best time to arrive at the volcano. The sunset and deepening darkness of the sky, and the gradually increasing glow and fire of the rocket-like bursts of red-hot stones and liquid lava, are scenes difficult to forget.

Valley after valley of this entire area have now-extinct small, medium, or large cones dotted over it, each one a silent, cold remnant of a former fire-spitting volcano. Once we counted twelve cones within view at one time, some of them old craters which now have cornfields cultivated in them by the Indios living nearby. And the road cuts show in almost every imaginable color the volcanic ashes of ancient eruptions deposited in thick beds.

Pine forests cover part of the country still and add much to its attractiveness. We ate lunch under a grove of lofty old trees with an acorn-eating woodpecker [Acorn Woodpecker], a new entry for my life list of birds, above us for company. We chewed and he chewed, very companionably, neither one of us molesting the other. The air was fresh and like wine at that altitude, and our appetites doubled as a result of it until we almost disgraced ourselves.

Then the country became more and more barren as we progressed, and the hills here around Lake Chapala are very dry. When we reached the lake, we turned off and put Chico into four-wheel drive. After forty-five minutes of driving first over a dirt road, second along an ox track, next through a farmer's field, and finally over open shrub-covered ground, we are now encamped fifty feet from the lake's edge in a lovely remote spot with scattered clumps of water locust trees around us.[1]

Since we plan to spend several days here and estimate in Chiquita all the waterfowl population on the lake, which is about sixty miles long and varies from between twelve and twenty miles in width, we are setting up a more permanent camp. Chiquita has been carried down to the water's edge. There we padlocked her with a chain around a tree trunk so that we don't have to carry her back and forth from camp as we did at Pátzcuaro, which we found backbreaking work.

The little nylon tent is up between two trees, and a large tarpaulin is spread from Chico's roof to trees behind him so that we have permanent shade for our work tables and camp chairs. The stove is set up, and the food box is out and under the truck. So we are already well established before nightfall.

Our first-sighted birds on the lake were a group of sleek black-necked stilts, boldly patterned in black and white, and not at all shy in this unfrequented spot. They have just walked back and forth at a sedate pace through the shallows before us seeking a supper of insects and other small fry in the water. So the camp is already named, and I have another new bird for my list. We will sleep well tonight as the little waves, pushed by a northwest wind, make a constant slapping sound against the shore.

January 11, Lake Chapala, Michoacán and Jalisco

We have had a wonderful day today. After hot oatmeal, lots of coffee, rolls and peanut butter at daybreak, we set out in Chiquita to estimate the waterfowl around the southeastern end of the lake. Today we saw coots, gadwalls, pintails, baldpates, canvasbacks, shovelers, scaup, blue-winged and cinnamon teal, ruddy ducks, and a few ring-necked and fulvous tree ducks [Fulvous Whistling-Ducks]. They flew or spattered off ahead of us in small clouds as we approached, with motor idling. Never have I seen such a sight before; it is really thrilling.[2]

At the delta end of the lake where the Río Lerma enters it, there is a large area of reeds and rushes. They have so grown in masses parallel to the shore as to form several waterways or lagoons apart from the lake proper. We found that this is a favorite spot for waterfowl.[3] And as we rounded each curve or plant clump, new groups were momentarily shocked into inaction directly in front of us. Then off they flew with a clatter.

The most memorable moment of the day to me was my first sight of the large western grebe, which looks not at all like its smaller pied-billed and Mexican [Least Grebe] relatives. Coming silently around a clump of tall rushes we saw Mother and Father grebe placidly riding the water, and near them three downy little gray grebe chicks floating slowly along like thistledown. For a few seconds, all remained as before. Then with astonishing speed and efficiency the little ones were gotten into a line and convoyed, like so many small freighters with the parents at the head and tail like battleships, in among the thick aquatic plants and out of our sight. And not another glimpse of them could we get. We were really very fortunate to see them at all since it is most unusual for them to have young ones so early in the season.

The boats used on Lake Chapala by the natives are an odd high-sided craft with long, pointed, covered prows. They have a more or less straight tree trunk for a mast, and a single large jib, the only sail, is rigged Marconi style. With such heavy planked boats their passage is naturally slow. And they

Most of the former vast delta of the Lerma and Duero rivers at the eastern end of Lake Chapala has been diked and drained for cropland. The channeled river is in the foreground; dikes separate it from the new agricultural land beyond. A few gleaming marshy spots in the fields are all that remain of what George Saunders described "as one of the finest waterfowl marshes in North America." The marsh originally reached to the base of the mountains in the background and for many miles up the river to the left. January 1948. (USFWS by George B. Saunders)

never seem to carry any oars, and probably couldn't move the boat much with them if they did. So when they are becalmed, they are becalmed indeed.

Each vessel keeps a tiny charcoal fire going in a metal container, and the men always have a supply of beans and tortillas and coffee. So whether it takes a few hours or two days to cross the lake, it is no very important matter. The heavy boats do have the advantage of standing the choppy waves, which the daily high wind produces, very well.

Two boats, sluggishly sailing along in company side by side, met us this morning and passed only a few yards away. As we skimmed lightly and rapidly along in Chiquita with the little motor humming smoothly, we thought that they looked at us rather enviously. But they all gave us a pleasant "Buenos dias," and a flash of the perfect white teeth which most campesinos have.

January 12, Lake Chapala, Michoacán and Jalisco

George went out alone today in the canoe to do more estimating, and I stayed near camp to collect several shore and other birds of interest for the museum. Little clusters of least and spotted sandpipers zigzagged irregularly through the very shallow water all day, prodding with their slender bills the sand before them with the unceasing rapidity of sewing machine needles.

Coots, only a few feet away from me, squawked raucously at encroaching competitors. Dowitchers [Long-billed Dowitchers], Wilson's snipe, glossy ibises [White-faced Ibises], and American egrets, as well as the black-necked stilts, appeared and disappeared at intervals. Through the shrubs and trees dozens of Audubon's warblers, goldfinches [American Goldfinches], blue grosbeaks, and field sparrows flitted and chirped. So today I have added two new birds to my life list, the Wilson's snipe and the glossy ibis, and a number of specimens for study skins for the museum's collection.

Chapala is an odd lake. Just as I think I have settled in my mind the exact colors of its waters, I look again and they are entirely different. Its depths are turbid with a yellowish silt and are never really transparent, so that the reflected colors superimposed on this give peculiar effects. Sometimes it is the mixed pastel blue-violet-pink of an old-fashioned flower garden. Again it will be heavy with intense mustard yellows and blue-blacks. And I have seen it the silvered gun-metal of a black-topped road after a shower of rain when the surface is still shining wet and one looks across it from a little distance.

It makes me feel after watching it that I shall never again be entirely satisfied with the simple innocuous blue of so many mountain lakes.

I was moved today to try my hand at a watercolor of it. And a maddening, frustrating experience it was. Having well started with it as it then seemed to me, I looked again a few minutes later and all the colors were entirely different. I was already having trouble trying to mix my paints to show the fuchsia, lavender, and chocolate colors it had started out with. So I finally had to give it up with only a poor, faint ghost of what the vivid scene had really been like.

While I cooked a hearty dinner late this afternoon, George skinned out and made up into museum skins the birds I collected today. Darkness comes very early here because of the mountains, and as soon as the sun goes down, it is bitingly cold. Our sleeping bags are really the only places in which we can keep warm. So we are going to bed each evening at sunset and having more sleep than we have ever had before since it isn't daylight until seven o'clock in the morning.

January 14, Lake Chapala, Michoacán and Jalisco

These are lovely early mornings with a light haze over the brassy-silver waters of the lake. The sun casts long golden rays over the deep blue-violet mountains on the far shore. And there is nothing to intrude on our enjoyment of the first cup of hot coffee and the first twitters of the sleepy birds around us.

We went out on the lake again yesterday and today in the canoe to observe and estimate waterfowl. But today such a strong wind arose about noon that we were shipping waves long before we could get back to camp and ashore. On the whole Chiquita rode them very well, and we were pleased with both her and the motor in their first real test. The motor hasn't caused us a moment of worry, for which we are more than thankful. Our past experiences with outboards have not been always so happy.

The hundreds of coots that cruise back and forth in front of our camp never fail to amuse me. They look so solemn and ministerial in their white forehead patches against glossy black plumage. And they bob their heads back and forth like chickens as they paddle along, rather in the disjointed way of too rapidly growing adolescents. They look the oddest though when startled and hurrying to fly away. Their bodies are apparently very heavy for their wings. So before they can achieve flight speed and leave the water, they must run along its surface for a long distance. This they do, frantically kicking against it at every step with their feet to help them get up to speed. Since their feet are not webbed like those of a duck, but have only flaps on the sides of their toes, they probably haven't too much paddling power. While this is going on, they twist their necks and roll their eyes and continually look behind them in a very droll manner to see what horrible danger they are escaping.

George has looked for meadowlarks all along the highway from east of Morelia to here without hearing or seeing a single one.[4] Although we passed dozens of very likely looking fields in which we felt sure that some smart meadowlark certainly should have homesteaded long ago.

This afternoon our patience was rewarded. South of the lake on a tremendous grassy pasture with cornfields nearby, we came upon a flock of several dozen. That is a major trouble in studying meadowlarks. They often are of very local distribution, and may be present on one quarter-mile stretch of fields, and absent from miles more all around the area that seem to be identical to the chosen one.

George did his imitation of a male's territorial song, and several answered him, apparently under the impression that he was a strange bird trespassing upon their already staked lands. Up they craned their necks, standing as high as possible on the top of grass tufts. We collected a number of them since George is anxious to get all the information on their races and distribution in the United States, Mexico, and Central America for his monograph on them. Also we ran into a new local name for them.

Several farmers were plowing nearby. So, following our usual custom, we took one of the collected birds to several of them separately and asked what he called it. They all responded with the same name, "Tortilla con chile," which seems a completely illogical one, and we can't imagine how it originated.[5]

Tortilla is the flat pancake of finely ground cornmeal that is the staple food for so many Mexican families in place of bread. And chile is a very hot pepper of which they often make a drippy sauce to eat on their tortillas. But the bird hasn't the color or pattern or taste of either of them. And it certainly occurs too sparsely and is too wary to be eaten commonly anyway. So we shall have to chalk up another mystery in the oddity of so many common names for both animals and plants.

They have a type of well at the small rural homes around the southern and western shores of Lake Chapala that we haven't happened to see anywhere else in Mexico. Beside the well there is embedded in the soil a high sturdy wooden post. Fastened to it there is a twenty-or-so-foot-long slender tree trunk as a crosspiece. On the top of the butt end of this crosspiece there is bound a large heavy rock as a counterbalance, and from its slender tip there dangles a bucket with a length of rope. As the rope is pulled down, the bucket lowers into the well and is filled. Then the heavy rock helps by its weight to raise it again to the top of the well. It is no doubt a primitive, but extremely ingenious labor-saving device.

We went down to the edge of the lake this evening to hear the flocks of mourning doves come in to drink. They always wait until the last vestige of twilight. Then suddenly they whistle down in converging groups from the hills in back of the lake which show dark in several shades of black. Visible only as startling silhouettes against a slightly lighter sky, they land on the extreme brink of the shore.

After drinking for perhaps three minutes, they all take off simultaneously again with a loud rush of wings to return to their roosts in the hills. It is an impressive sight and one which always makes a shiver run up my spine as they without warning noisily volplane in only a few feet over our heads as we stand quietly hidden in some clump of small trees.

January 15, Lake Chapala, Michoacán and Jalisco

Today, Sunday, has been visitors' day at camp. We had planned to spend it writing up notes and necessary memos, and doing office work in peace and quiet. But the passersby and the noise of our motor on the lake have gradually spread the word among the locals that strangers are stopping here. A little boy with a small flock of goats opened the day when he stopped and examined us critically as we were breakfasting this morning. In the meantime his omnivorous beasts ate my favorite clump of trees on which I have been hanging the dishtowels to dry.

Shortly afterward a group of women riding undersized, mouse-gray burros came along and stopped nearby while their offspring cut wood to carry home. Then half a dozen of the young bloods from the nearest village rode over on horseback and proceeded to put their steeds through their paces as boys will do anywhere to show off. Having nothing better to do, they stayed for hours watching our every move from only a few feet away until we were thoroughly tired of having them almost at our elbows. Had they all been well behaved and friendly, we wouldn't have minded having them nearby. But, unfortunately, they became more and more boisterous as the afternoon advanced.

Our favorite of all the visitors were a little old man and his wife, apparently small farm owners, who dressed up in their cleanest and neatest white cotton clothes and walked over to see us. Both were so elderly they were shriveled and tiny in size. But their brown eyes gleamed bright as children's with interest in all they saw, as they stopped a polite distance away and looked at everything we had and were doing.

They had a spirit of companionship as they commented and chuckled together upon things that amused them that we have not seen before among the rural folk. Usually there is little or no conversation as such between husband and wife. The man is lord and master and tells the woman what to do. And she obeys, even to the extent of following in public a correct several paces behind him like any well-trained dog or horse is expected to do.

We talked a little with the old couple and told them why we are here and what we are doing. They showed an intelligent interest, and we enjoyed them thoroughly until they felt it was time to go. Then with a proper, "Adiós señor, adiós señora," from each in turn off they went back to their farm. They were probably full of tales to tell their neighbors and relatives about the odd things Americans are moved to do and the incomprehensible stuff they lug about with them.

January 16, Guadalajara, Jalisco

We drove on along the sixty-mile length of the lake early this morning after breaking camp. To us, Lake Pátzcuaro is much more beautiful because of its finer setting, the higher mountains around it, and the more varied and interesting islands in it. But Chapala in a different way has beauty too. It is long and narrow, and several villages with red tiled roofs can be seen at once on opposite shores. And along its surrounding mountains the lilacs deepen into the velvety violet that seem always to be present along the flanks of the mountain landscapes of Mexico.

At the far end of the lake we turned off to the right from the main highway to Guadalajara and onto a rough dirt road that leads around the north side of the lake to the town of Chapala. The trip was slow, but very interesting as we passed through a number of little, usually bypassed villages. In the first one, Jocotepec, a group of dancers, togged out in brilliant cerise and gold costumes and masked, were preparing to dance in front of the church in honor of some local feast day.

The towns and churches around the lake are old, and the latter may well date from early Spanish colonial days. The towns undoubtedly antedate the coming of the Spaniards by several hundred years, since Jalisco is thought to have been settled in the tenth century by Nahoas.

Most of the villages along the lake still have the same names they had before the Conquest. But many of them are now preceded by a saint's name which was tacked on by the Catholic Church, apparently with the idea that the good influence of the latter would certainly sanctify and counter-balance the paganism of the former. Thus little Cosalá now glories in the title of San Juan Cosalá, and may or may not be any more holy as a result.

We finally bounced through little Ajijic, which is enjoying a mild boom as the chosen settlement of a group of American artists and writers. It is perhaps the only place in the world with four consecutively dotted letters in its name, which may possibly have something to do with its attraction for them.

Then we drove on for a few more kilometers to the town of Chapala. Since the two main streets in Chapala are almost entirely torn up in an effort to widen and modernize the place, our impression of it is perhaps unduly unfortunate. Also, the water of the lake, on the edge of which Chapala is situated, is going down so rapidly that it will soon be left high and dry. Even now bathers must wade out a hundred yards or so to be in waist-high water. We had an excellent lunch in a large open-air restaurant directly on the beach, and I enjoyed once again my two favorites, *pescado blanco,* white fish, and *pollo en mole,* chicken in *mole* sauce.

After lunch we started up the formerly paved road from Chapala to Guadalajara since there is a series of large reservoirs and small lakes which George wanted to scout for waterfowl. This road too is undergoing reconstruction, so it is much rougher now than it has been in the past. By taking side dirt roads, we managed to reach half a dozen of the larger reservoirs and lakes we had planned to visit. George made his estimates satisfactorily, and we rolled into Guadalajara shortly before dark, the country becoming more and more arid and barren as we advanced.

Guadalajara lies on a fifty-two-hundred-foot high beige-colored, dusty plain and not among forested mountains as I somehow always expect this second-largest city in the republic to be. Its praises are usually sung in Mexico for its antiquity, charm, and beauty, but we have tried in vain for this impression. Just now an extensive rebuilding program is going on among the central blocks of the city, and unfortunately some of the most interesting things about it, the picturesque old portales or arcades around the Plaza de Armas, have been razed.

We are at a comfortable hotel on the outskirts of town which has a pleasant garden where George can spend several days doing some of his necessary office work, and Chico can be safely parked off the street in a patio. It has been terribly cold the last few days, and the headlines in today's paper mentions a *fuerte tormenta* in Veracruz. So a "norther" must have come down from Texas, and we are within its range. As in Florida and Southern California, sometimes in the winter hotels and houses that are built only for hot weather with no artificial heat can be pretty miserable when unexpected cold arrives.

January 19, Guadalajara, Jalisco

Guadalajara was founded by the Spanish in the 1520s. The Guadalajara people are officially called Guadalajareños, but are generally spoken of as Tapatíos. This is probably from their habit in the early days of wearing the heavy embroidered Aztec shawls called *tápalos*. A number of the inhabitants today still trace their lineage in a direct line from the early conquistadores and the Spanish royal houses. And many people consider it the most Spanish city still in Mexico.

One local custom, which probably no tourist misses if his stay is longer than overnight, is that of having wandering bands of singer-musicians, known as mariachis, present for every possible occasion. They are encountered everywhere in the city . . . in front of bars, on the sidewalks, in restaurants, in private homes for birthday and wedding celebrations, and of course at the tourist hotels. Some of them are excellent and a delight to listen to. And some of them are terrible.

These roaming, guitar-playing singers apparently originated in Guadalajara, although they are now common all over Jalisco, and are even seen as far away as Mexico City. Their name is thought to come from the French, who were also early residents in the city, and their custom of always having singers at their weddings. The French word *mariage* could, without too much difficulty, have been transposed by non-French speaking people to *mariachi*.

There are several very fine frescoes in the city, painted by José Clemente Orozco, which are well worth seeing. I went today to see one that he painted in the Palacio de Gobierno, the "Palace of Government," which faces the Plaza de Armas from the east side.

This fresco covers about four hundred square feet of surface and is unusual in part because it is painted on the walls of the main two-story staircase and on its dome above. It is an affair of bold lines and vivid raw colors and represents man, and his progress and philosophies in the material and spiritual worlds. Parts of it will shock the average viewer out of his customary apathy because of its faithful representation of the less desirable sides of these philosophies and progress. And I believe that no one will forget it for some time.

To me the most striking part of it, and from which I could hardly tear my eyes away, is the great central figure of the Mexican patriot Miguel Hidalgo y Costilla, who actually wrote part of his Declaration of Independence in the palace in 1810. It is a tremendously powerful and violent conception of a man and completely dominates the entire painting.

January 20, Guadalajara, Jalisco

We have taken a field trip today to estimate the duck population on two lakes which are shown on our U.S. Army Air Corps maps of this section to be southwest of Guadalajara. We drove back along the main Guadalajara–Mexico City road and turned off from it to the west, passing several large sugarcane fields and a few small villages. Then the countryside became drier and drier.

Soon we were following a single car track through loose, light yellow soil that billowed in clouds as we passed, with the usual desert types of plants around us. We saw almost no one except a man with a burro, and an overloaded bus going we couldn't imagine where.

Finally it was a question of simply taking off cross-country to where we thought the lake should be. Chico was equal to the occasion, and before too long we were within sight of the lake, which is locally called the Playa de Atotonilco. Nothing but cacti and thorny shrubs in the way of plant life were to be seen, except at the far side of the lake from us where a group of dark green shade trees indicated the site of a small village.

George estimated several hundred ducks to be on the lake as a whole, including gadwalls, Mexican ducks, shovelers, and blue-winged teal. When he had finished his work, we had a sandwich lunch under the shade of a giant candelabrum cactus and then started out again for the second lake, which is called the Playa de Zacoalco on our map. An hour's drive brought us to and beyond where we were sure it should be located from the mountains and other topographical features of the map.

But no lake appeared. Finally we were positive we must be past it, so we turned around and were soon lucky enough to see a man on horseback at some distance from us on a flat plain we had previously crossed. We bounced over the rutted hardpan to him and discovered that at that moment we were in, or rather on, the lake. Except that there isn't any lake there in the dry season. The rains form a shallow sheet of water for only a few months each year, which is the playa of our map when it's there. So feeling quite cheated, we made our dusty way back to the main road and to Guadalajara.

However, during the course of the day even in that dry barren country, I garnered a total of five new birds for my life list. The first was a fine big Sennett's white-tailed hawk [White-tailed Hawk], which sat stolidly on the top branch of a dead tree and let me peer at him for as long as I would. Next were two flashing male golden-fronted woodpeckers that lack gold on their fronts at all as I would interpret it. Their breasts haven't a bit of gold on them, and what there is, is on their foreheads above their beaks. Thirdly an [American] avocet stepped off a measured pace along the edge of the playa as we ate lunch. Fourth, a blue-gray gnatcatcher repeated incessantly his mouse-like squeak; and lastly, a tiny Sharpe's seedeater [White-collared Seedeater] had an early supper in front of us along the main road on our return trip.

CHAPTER 4

Mazatlán and Tepic

January 22, Mazatlán, Sinaloa

GEORGE IS through with his fieldwork of the annual waterfowl inventory and has only to write his report on the results to send in to Washington. We are now going to continue with his regular biological research on the duck wintering grounds in Mexico. So the next place to visit is Mazatlán in the state of Sinaloa on the Pacific coast.

This is about 340 miles northwest of Guadalajara, and George wants to study the lagoons to the north and south near the small city. He has been there twice before, but only by air. Each of the two winters when he flew entirely around the coasts of Guatemala and Mexico by small plane, it was an overnight stop, and he liked the town and people very much both times. Mazatl is the ancient Náhuatl name of the early Indio inhabitants for deer, so Mazatlán means literally, "Place of the Deer."

We started out today after an early breakfast to drive there and soon passed along the outskirts of the town of Tequila. This little village has given its name to an alcoholic drink which is produced in the region from the sap of a smaller agave than the maguey from which pulque comes. There were tidy rows of the spiky blue-green plants, very attractive to the eye, for mile after mile on both sides of the road.

Farther along we noticed huge chunks of black stone embedded in the ashy, volcanic soil. A fresh road cut at one place finally made us realize that what we were looking at was obsidian. It is that peculiar black, glassy mineral which the aborigines of many volcanic countries used before metal was discovered to make their sharp-edged weapons and kitchen utensils.

We have both seen many beautiful examples of this handicraft in the museums of several countries. And I have had the thrill of digging up a number of these small, two-edged ancient knives of the material from my own garden in Guatemala. But neither of us has ever before heard or seen where it came from in Mexico in bulk form. As we prepared to park Chico off the road, we came upon the largest chunk of all, some six by six feet, with many smaller pieces around it.

Out came our geology pick, and we chipped off samples for ourselves and interested friends. But we still don't know how the early residents managed to make the beautiful knives with their flat, perfectly straight ridges down the center and their sharp chipped edges. No matter from what angle I start, nor how carefully I tap it, my pieces always shatter into perfectly useless fragments.

Farther on a great escarpment appeared directly before us, and down we crawled, with Chico in second gear, to the river valley below. It was an odd experience to drop so suddenly from the arid highland plain to a lush, semitropical vegetation of sugarcane, banana plants, and papaya trees at the bottom.

When we were almost down the great slope, I happened to see from the car window what looked like an unknown moss growing in thick, blackish clusters against the ochre walls of the road cut. George found a safe place a hundred yards beyond to park Chico off the road. And I started back with a small paper bag and a knife to pry a good sample of the moss from the rocks and dry it for an herbarium specimen.

When a yard away and opening the sack, I thought the moss seemed oddly spotted in places. But not until I bent down a few inches away, with my knife point touching one edge of a thick black cluster and it suddenly dissolved before me, did it dawn on me what it really was. Hundreds and hundreds of daddy-long-legs were tightly packed together. Their skinny black legs resembled the thick moss, and their tiny pale round bodies, only a few of which showed clearly through the higher legs, made the oddly colored spots I had noticed. They flowed rapidly away and scattered as the impulse from the one knife touch at the side of the clump was communicated through them. Each group was in a slight natural hollow which had enough of an overhang to form shade over them during the heat of the day.

We continued on our way, mossless, and made a brief detour to estimate the waterfowl on Lake Magdalena near the town of the same name. The lake turned out to be smaller than we had thought from the map, but a hundred or so [Greater] white-fronted geese on its surface made a fascinating new bird for my life list. And there were several hundred ducks of various kinds.

Soon we started out again and farther on drove through a thick old lava flow which long before had spilled over the edge of a volcano that still stands five hundred feet or so to one side of the road. The highway has been cut directly through it so that the forbidding ugly lava rises up above one's car on each side in jagged black ridges.

We had already entered the state of Nayarit, a new Mexican state for me, when we met three campesinos in their shapeless white, handwoven cotton clothes of a style we have not seen before. They were apparently down from some very remote region since the man carried short bows and arrows, and they all wore their hair long and falling over their shoulders.

Two had trousers and one a skirt, so we assumed they were two men and a woman, although their hairdos were identical and we could not tell any other way. They trudged stolidly along the dusty shoulder of the road, looking nowhere except straight ahead. We couldn't help wondering how many days in all they would have to spend walking before they were safely home again.

The road turned out to be paved as far as Tepic and beyond it down the long steep winding mountain to the coastal plain. Then it became rough gravel with occasional *chalanes*. *Chalanes* are local ferries where one drives onto a small one-car (usually) barge in solitary splendor and is pulled across wide unbridged rivers . . . usually by man power. For some seventy miles the gravel road lasted.

The part that really slowed and jolted us up the most was the innumerable *desviaciones* or detours around the dozens of washes and arroyos which the highway department is bridging. We were thankful to have an older-style car with high clearance and not a modern low-slung one, as we would certainly have been hung up several times on rocks where the road-building equipment had piled gravel or scraped long rows of soil.

Shortly before we reached Escuinapa, the paved road started again. We went along at a good speed over it until we reached the outskirts of Rosario and hit, by particularly unfortunate timing, a local repair job along the main street on the bridge over the river. This turned us off onto a long, winding, dusty detour. We also had to eventually cross the rather wide river by means of a ford, and the water was so high that it came in on Chico's motor and stopped him cold.

There we sat marooned in mid-stream just at the last light of evening with no one in sight but a couple of burros who thoughtfully regarded our plight. I had visions of spending a cramped night curled up on half of the seat. But eventually we got Chico going again on a couple of cylinders and by keeping them going spasmodically, we dried off the wet ones and eventually crawled out.

An hour and a half after dark we rolled into Mazatlán. We are weary, both entirely covered with a thick layer of sticky yellow dust, and jounced until our back teeth ache, since no truck's springs are too comfortable even on a good road. After hot showers and plenty of substantial food, we are now more than ready for bed.

January 23, Mazatlán, Sinaloa

We fell into bed last night and more than slept the clock around. Today I can appreciate, as I couldn't last night, the beauty of the town's surroundings. We are in a pleasant modern hotel directly on a small horseshoe beach, and there are even finer, larger cream-colored sandy beaches to the north and south of town. Several precipitous rocky islets off shore add to the beauty of the view. The water is only medium cold and quite calm, and not so rough as the Pacific Ocean is usually along California or South America. So good swimming can be enjoyed the year around.

I have walked all over the city today, having a dozen or so necessary things to buy, and I am entirely charmed by the attitude of the people. They are unfailingly polite and most helpful. If they did not carry what I asked for in a store, as was several times the case today, each without exception told me just who did stock it and pointed out the way. All over town, men with little pushcarts sell fresh oysters, opening them as one picks them out and serving them with limes, catsup, and Tabasco sauce. They are cheap and delicious.

Several hours were spent in the large central market with its many and varied stalls. In one section devoted to handwoven goods, I bought a many-colored shopping bag woven from the fiber called *ixtle,* which is obtained from the leaves of agave plants. It is sturdy and just the thing to hold our lunch makings in the back of Chico.

Some of the vendors in the market have as pets small green parrots [probably Barred Parakeets] of a type new to me. Their main feature is a large circle of bare yellow skin around each eye. This makes them all look like miniature Harold Lloyds and gives them the slightly jaundiced men-of-the-world air which Walt Disney's parrot, Joe Carioca, had in *Saludos Amigos.*

They are amiable and very friendly little birds, but do little or no talking. The woman from whom I bought some very excellent tangerines had one, and seeing that I was interested in it, she took it out of its cage which hung over her stall.

The little fellow sat quietly on her forefinger regarding me owlishly while she stroked it. According to her they commonly live along this part of the Pacific coast and are frequently caught when young and sold as caged birds. She said it ate anything she sold in her little section and everything it could get from those of her neighbors. From the variety of goods displayed, it must have the digestion of the proverbial horse.

January 26, Mazatlán, Sinaloa

We have spent the past three days on the lagoons, both in Chiquita and wading in rubber boots. Our main occupations during that time have been taking salinity tests of the water, samples of the bottom soils, and collections of foods upon which the birds are feeding. The days have been interesting and the evenings superb with brilliantly colored sunsets.

There are several little lagoons within easy driving distance from town, and we have also seen a fine lot of sunrises from them, as it has been necessary to start our studies at an early hour. Small flocks of lesser scaup, ruddy ducks, and pied-billed grebes have been our constant companions.[1] And large numbers of avocets and willets have daily paced along the edges of the shallow water, dining as they went.

George did some duck collecting in order to examine their stomach contents and to see what plants and mollusks present there they are using for food, and in what proportions. We were greatly surprised to find that the

A small portion of the huge Marismas Nacionales south of Mazatlán, Sinaloa, in 1948. These saline and brackish Pacific coastal marshes rise and fall in accordance with the annual rainfall that begins in July and provide excellent habitat for many species of waterbirds, including ducks, throughout the winter. They are generally considered to be the best of the Mexican Pacific coast marshes. (USFWS by George B. Saunders)

Mazatlán, Sinaloa,.as it appeared in January 1948. George no doubt scouted the lagoon for wintering waterfowl. Today the area shown is dominated by hotels and other development. (George B. Saunders files)

gullets of a number of them were simply crammed with thousands of small flat organisms. George thinks they are the larvae of midges which are commonly called blood-worms, although they aren't worms at all.

There is a beautiful walk of about a mile around a rocky headland on the edge of the sea and out to a former islet, now connected to the mainland by a causeway. At the farthest point of this islet there is a lighthouse. Standing as it does 515 feet above the water, and with a light that is visible for thirty miles, it is one of the highest in the world. There really isn't anything for a tourist to "do" here in the sense of actively dashing around to various sights or places, except for the deep-sea fishing, which is excellent. But it is one of the best places I know to vegetate and do nothing pleasantly.

The air is balmy. Each day is sunny and blue-skied like the one before it. And there is an air of tranquility that is very conducive to the growth of enormous appetites. An American who has lived here for many years has a restaurant near the two hotels on the beach drive. Here excellent hotcakes and syrup, ice cream made of Borden's cream, and many other splendid dishes such as lobster and shrimp are served. So there is every reason to enjoy a large appetite and a pleasant stay. But although we would both love to stay much longer, tomorrow we must be on our return trip to Guadalajara.

January 27, Tepic, Nayarit

We left Mazatlán early this morning and turned off on a dirt road at the town of Escuinapa, about fifty-five miles along the way, so that George could do some waterfowl observing on the sea marshes nearby. That completed, we returned to the main highway and were soon again on the seventy miles of gravel road. By the time we had jounced very many miles, our dislike of the word *desviación* was amounting almost to a couple of full-fledged fixations.

But from the last *chalán*, the remaining drive up the mountains to Tepic was sheer pleasure. Although it was a slow trip since by this time the jeep was affected by the rough road. The caster and camber of the front wheels are off from the constant jolting, so we had to make the last part of the drive at a very slow speed to keep from having them shimmy so violently that we would have to come to a dead stop.

We turned off from the bypass around Tepic and drove into the city proper, as it was almost dark, and we can't possibly make the 140 miles on into Guadalajara anyway until the wheels are adjusted. It is just as well, as George spotted some meadowlarks just outside the city and is most anxious to find out to what race they belong.

Having looked for a suitable hotel in central Tepic and finding only overpriced and dingy rooms there, we fortunately remembered a conversation we had a few weeks ago with an American couple who had recently spent a week in Tepic. They had strongly recommended the Hotel J., a small but modern and clean place with a dining room, on the Calle Juárez farther out from the center of town and nearer the bypass. So we tried there and are now comfortably settled in it. The owners speak only Spanish and are as Mexican as tortillas and chile except for two things. They have surprisingly dark blue eyes, and their name is straight out of Ireland, legacies from some distant ancestor.

Just at dusk this evening while we were in our hotel room, we heard several hoarse croaking calls that could only mean night herons [Black-crowned Night-Herons]. Not expecting to see them up here at 3,188 feet in altitude, we both rushed for the windows, disbelieving our ears. Yet there they were, half a dozen, in a long slender line. Their thick bills were extended before them, and they were silhouetted against the darkening sky like a beautiful animated Japanese print. They were doubtless flying out from their daytime roosts to feed.

The air is pleasantly cool, and we have enjoyed a leisurely walk through the nearby part of town this evening. Now that we have encountered considerate treatment at reasonable prices, our attitude has changed and we are inclined to like Tepic in general. We have just been discussing how such comparatively trivial occurrences can permanently color one's opinion of a place forever after. We might easily have left disliking the place heartily, and of course telling our friends about it in that light.

January 28, Tepic, Nayarit

We certainly received royal treatment at our simple little hotel this morning at breakfast time. In the dining room last night at dinner the owners were greatly distressed because when we asked for fresh fruit for dessert, they had none. The Señora said, "Wait until tomorrow and we will have everything you want." We thought it all too likely that it was just a figure of speech and meant nothing.

Our surprise was great this morning when we ordered fried eggs, black beans, rolls, and coffee with hot milk (our standard breakfast when in small hotels as there isn't much a cook can do to ruin it), to have brought on large glasses of chilled orange juice, slices of papaya with half limes, big dishes of fresh strawberries covered with whipped cream, and an entire hand of big ripe yellow bananas!

This was filling our wants with a vengeance. We ate all of these delicious extras we could possibly swallow, besides the regular breakfast we had ordered, and our total bill for all of it came to just forty cents (American) apiece. I must say it is the first time such an experience has come my way, either in Mexico or the United States, or anywhere else. It isn't typical of Mexico by any means to find such an amount so kindly served for so little, nor would it be in the United States. But it shows it *can* happen.

George took the jeep around to a garage the first thing this morning to have the wheels realigned, and in a few hours it was fixed and ready to test run. So we decided to go out to where we saw the meadowlarks yesterday and do a little collecting for their better identification.

Just as we started at one end of the field to stalk two male meadowlarks which had recently sung, one of the most unusual sights I have ever seen occurred. A low indescribable sound started, not very far away. It seemed to be partly a whirring noise, partly low thunder, and partly the high sustained note of rushing, turbulent air. Just at ground level there soon appeared, moving at a constant speed, an amorphous black mass that flowed smoothly up and over all obstacles it approached. It cleared them by not more than a foot or two, and constantly changed its shape like an amoeba putting out pseudopodia, or false feet. But all the time it remained a tightly packed body with definite and distinct outlines.

I couldn't imagine what it was. It turned out to be a gigantic flock of yellow-headed blackbirds, thousands upon thousands of them. George estimated there were over twenty thousand birds, all moved by some unknown impulse to band and fly together. Within the margins of the mass individual birds constantly soared or dropped, and positions were frequently exchanged so that they appeared to be dancing black motes against the pale blue of the sky.

The noisy whir of their wings frightened the meadowlarks in the field, and off they rose to fly before the dark cloud for a long distance until one by one they dropped down into the grass again. The blackbirds went on out of our sight still arranged just as compactly and flowing just as liquidly as when we first saw them. It seems incredible that so many could fly together without constantly knocking down neighbors with the flapping of their wings.

We had good luck with our collecting eventually and obtained a respectable series of two races of meadowlarks, the western and the gold-breasted. George is particularly elated about the latter, as he separated and named this race some years ago from material in the Chicago Natural History Museum because of its distinctly saffron-yellow breast and the different coloration of the tail feathers and back. This is the first time it has been found in the state of Nayarit, so it enlarges its range in Mexico.[2]

But an equally great and more unexpected prize was the obtaining of two male white-winged doves, a bird species he has studied for years and years.

These two specimens are both members of the western form and are the first we have seen this year in Mexico, and also the first in Nayarit. Although, of course, we have looked for them wherever we were.[3]

We had a sandwich lunch in the field and returned to the hotel about four o'clock. Late afternoon and this evening have been spent in preparing the birds as museum skins for later study, and the writing up of notes on weight, sex, parasites, and other important taxonomic characteristics. We have had another fine dinner and are now ready to tumble into bed, as we are still tired from yesterday's drive.

January 29, Tepic, Nayarit

This morning George and I got up at six o'clock and drove out about eight miles from town to the same place where we found the two male western white-winged doves yesterday. He is very anxious to obtain a female or two from there to complete his studies. So we located ourselves very unobtrusively in high dead grass with two trees and a small stream of water directly before us where they should come to drink.

For years George has wanted to have an accurately transcribed musical score of the lovely songs of the whitewing. Some of these are several times longer and vary greatly from the usual short one which is commonly interpreted as "Who cooks for you?" in English and *Qué triste estoy*, "How sad I am," in Spanish. But to do this transcribing accurately, one must have a musical instrument of known key and portable size at the location of the song. And then it must be quickly figured and fingered out, note by note, as the bird repeats the melody. In the past, playing no small instrument himself, it has not been possible for him to obtain these unknown facts about them.

This year, however, for just such a contingency, I brought along my recorder, the original type of vertical flute with which Mozart and Johann Sebastian Bach were composing when they wrote *Flauta* on their scores. And in praise of which both Shakespeare and Samuel Pepys wrote. Since George excitedly announced yesterday that for the first time in Mexico this year he had heard male whitewings singing, I of course brought the recorder and a notebook along this morning to try and get some of the songs down on paper.

The sun had just risen above the top of the extinct volcano to the east as we got ourselves settled. The air was crisp and cold, and everything seemed clean and fresh. While we waited an hour or more for the first whitewings to venture down to the pool to drink, I had time to figure out and write down six different songs sung by six males perched in trees in the neighborhood. In all of them there is a range used of only two or three notes, and the variations consist in (1) either different groupings of these notes, or (2) in the same groups being rearranged in order, or (3) in a very considerable difference in the length of time the individual notes are held during a song.

Each variation so far has been repeated over and over without change during any one singing by a bird. The next outpouring may be the same one, or another. But so far none of the comparatively few I have been able to record has sung in the longer songs any one of another bird's variations.

The morning was a beautiful one as it slowly warmed, and the small birdlife was abundant. As I sat quietly, with George and his gun a hundred yards away, bird after bird intent upon an early breakfast, fluttered or perched within two or three feet of me before realizing that I was there. I gave a nasty shock to a ruby-crowned kinglet that perched on a stem directly by my head before it saw me. Then he immediately became half of his former midget size as he flattened his feathers down in fear and bolted off in a rush.

Two hummingbirds contested at intervals the rightful ownership of the little stream and a nearby flowering shrub, and fluttered up and down as they chittered shrilly at each other and battled briefly over the matter. One put on a show that I shall never forget.

The sun was directly behind him so that as he slowly hovered with wings and tail feathers widespread, he seemed to be entirely outlined in a moving halo of light. Up and down he effortlessly floated in and out of a small group of tiny insects, which themselves seemed as though made of pure silver as they danced against the sun. After each deliberate snap of his long coral-red bill, there was one less mote to be seen. And on and on he went like a ballet dancer until none was left.

Half a dozen kinds of sparrows, dickcissels, pileolated [Wilson's] and myrtle [Yellow-rumped] warblers, Mexican [Lesser] goldfinches, assorted flycatchers, blue grosbeaks, shrikes, [Northern] mockingbirds, and little Mexican ground doves [Common Ground-Doves] flew in and out of the trees and grasses and splashed in and out of the stream as they drank and bathed. They were either unknowing or unconcerned with the fact that humans were so near. Suddenly without the slightest sound there were two whitewings in the tree nearer George. He collected a male and a little later two females, so he can now check on the facts he wants to know.

Fig. 1. Transcriptions of songs of the White-winged Dove.

A. *Z. a. mearnsi* No. 1. January 5, 1949. Wild bird at Playa de Coyuca, near Acapulco, Guerrero.
B. *Z. a. asiatica* No. 1. December 11, 1949. In cage at Hotel Balneario del Río Pánuco, Tampico, Tamaulipas.
C. *Z. a. asiatica* No. 1. Same date and place as above. This is the common 'short' song, generally phrased as "Who cooks for you?", or "*Que triste estoy*" ("How sad I am"), which all white-wings sing.
D. *Z. a. asiatica* No. 1. December 13, 1949. Same place as B.
E. *Z. a. asiatica* No. 2. Same date and place as B.
F. *Z. a. asiatica* No. 2. December 16, 1949. Same place as B. This song was given at 11 p.m., the lights near the bird cages in the hotel being on at that hour.
G. *Z. a. mearnsi* No. 2. February 12, 1950. Ten miles south of Tepic, Nayarit, at 3200 feet elevation.
H. *Z. a. mearnsi* No. 3. Same date and place as G.
I. *Z. a. mearnsi* No. 4. Same date and place as G.
J. *Z. a. mearnsi* No. 5. Same date and place as G.

Musical score of various calls of white-winged doves recorded by Dorothy Saunders during the 1948–49 field trip. (Saunders 1951)

Then we sat together and watched the birdlife through our binoculars and enjoyed the warmth of the sun and the quiet of the countryside for a while before starting to do some general bird collecting for the museum. Of the nine specimens we chose to take between us, the most interesting is a magnificent little bird which neither of us has ever seen before, although we think that it is a tanager [probably an Elegant Euphonia]. It is large sparrow size and has an almost unbelievable combination of colors. Its head is powder blue; the face, back, and tail are a rich Prussian blue; there is a forehead patch of deep gold; and the breast and underparts are burnt orange.

Another unknown is a very small brown woodpecker [probably a Strickland's Woodpecker] about the size of a large wren. It is boldly speckled with white and has a small but flaming red crest, a broad white eye streak, dark brown cheeks, and the stiff bristly tail feathers that help support him against tree trunks in his perpetual search for insects.

The third unknown of the lot we think is a silky flycatcher [probably a Gray Silky-flycatcher], although we must still check to be sure. Long, very soft, light gray feathers cover its head, back, and breast. Its tail is black with a wide white band across its center, the abdomen is white, and the flanks and undertail coverts are a bright yellow.

All in all we are very pleased with the collection, and they should make up into very useful and interesting skins. Finally our breakfastless stomachs forced us to return to the hotel. After all it *is* Sunday and theoretically a day of rest. And it will take all afternoon and most of the evening to prepare the birds and to write up the necessary notes about them.

January 30, Guadalajara, Jalisco

We really hated to leave Tepic this morning, but unfortunately there are no waterfowl nearby in the winter, and the work on the other birds is finished, so there is no more reason to stay.

We were sent off from the hotel with a package, a gift from the owner, of six of the delicious fresh pork tamales that the cook there makes. Each one was only a few hours old and still wrapped in its protective folded corn husk. Latin Americans in general look with suspicion upon picnic lunches of sandwiches. And our friends at the hotel were sure that without their tamales we would not have a suitably sufficient lunch when we stopped along the way to eat this noon.

The first twenty-five to thirty miles south of Tepic are through beautiful country. Much of it is still wooded with pine and oak forests dotted here and there with the cones of extinct volcanoes. As we neared Ahuacatlán, an especially large volcano reared high above and directly to the left of our road.

We were looking at it idly and commenting upon its age and the knife-like edges of the tremendous flows when I was startled to see a bit of white steam puff up from one side of it. When we passed here just eight days ago on our way to Mazatlán, we saw no sign of any smoke anywhere. So as we drove today along its base, we looked carefully all over the old giant and were rewarded by seeing at least two dozen "smokes" bursting forth from its crinkled seams.

It has apparently been a long time since the volcano was active, and I hope that this isn't the prelude to another outpouring. There are quite a few prosperous little farms around the base, and they would certainly be badly harmed if not entirely destroyed by another eruption.

We continued along our way and again made the two steep drops through fiercely eroded barrancas (deep gorges) to little subtropical valleys at their bottoms. Then we huffed and puffed and climbed again to Guadalajara's dusty arid plain, and eventually pulled into the patio of our hotel.

George went to the Consulate for mail this afternoon and to the railroad station for tickets. The road from Guadalajara to Manzanillo down on the Pacific coast is not yet completed, and the big lagoon just south of Manzanillo is one which he must study. So we plan to leave Chico here in storage and be off tomorrow by railroad for a few days of work there.

CHAPTER 5

Manzanillo

January 31, Manzanillo, Colima

OUR TRAIN left on time this morning at a little after ten o'clock. For several hours we moved along an arid beige-colored plain which has a few small ponds or reservoirs scattered over it and two medium-sized lakes. Then for almost an hour we skirted what at one time was a very large lake, but which is now entirely a dry lake bed filled with powder-fine yellow dust.

At shortly after 2 P.M. we passed through Ciudad Guzmán, and from there on to Colima the scenery became really striking. The mountain peak of El Nevado de Colima with its attendant volcano, El Volcán de Colima, which juts off from one side and is only a little lower in height, dominated the landscape for hours with its altitude of 14,370 feet.

We followed above the deep roaring gorges of the Río El Naranjo for many miles, occasionally crossing side tributaries or branches on high suspension bridges, which gave brief shuddery glimpses of unknown depths below. As we gradually lost altitude, the vegetation became more and more lush and tropical along the rocky sides of the river, and more and more interesting to me.

At Tuxpan the entire train, first- and second-class coaches alike, was without warning deluged with hundreds of natives. They crowded into the aisles until literally no more could force themselves aboard. And there they patiently stood for miles and miles, literally packed like sardines. They were neat and clean and quiet, and the women wore a costume I have not seen before.

The skirt is of handwoven dark blue wool and so full that the extra material is folded back and forth at the rear of the wearer to make a six- or seven-thickness flat pleat, about a foot wide. This makes a permanent soft cushion to sit on, as the entire skirt is held in this form by an intricately woven narrow red, black, and white belt around the waist. Above, a handwoven blouse of very fine white cotton material, which has a narrow piping of either dark blue or red up the sides of the sleeves, is tucked into the skirt. Hair is worn long and braided, and feet are bare. The men wear the more usual pajama-style white cotton trousers and loose shirt with a broad-rimmed straw sombrero, and usually leather sandals.

At some unidentified, to me, wayside point the train stopped, and they all descended as suddenly as they had come to attend a fiesta in a nearby pueblo. We continued the trip in relatively spacious comfort. By the time we reached the city of Colima, we had been in the state of the same name for an hour or so. It is a new one for me, although George has been here several times before when he was making the annual waterfowl inventory along the coastline. At Colima we were down to about fifteen hundred feet in altitude. A half-hour stop let us see the various native types and the vendors of many foods who were there to meet the train.

Continuing, we gradually lost our remaining altitude, but unfortunately arrived at the coast after dark. Then we turned north and ran for twenty-five miles or so along the narrow land ridge between the sea and the Laguna de Cuyutlán, which George has come to study. At the end, the train suddenly and unexpectedly reversed and then backed for a mile or two on a causeway across the lagoon. About 9 P.M. we were all at once in Manzanillo.

It was a beautiful sight when we arrived. The city looked very much like some of the Italian towns which are built on the steep cliffs above the Mediterranean Sea. The lights gleamed brightly, row above row, and were reflected to sparkle again in the waters of the lagoon. The precipitous little hills stood out as black masses against the stars.

The town is small. Everything is within walking distance of everything else, and we are very comfortably settled in a hotel room directly overlooking the bay. I feel a little as though I were in a goldfish bowl, as our room is rounded with two sides made up almost entirely of glass windows. The cross-ventilation is wonderful, as it is hot and humid in the streets.

Below us in the little flower- and tree-filled plaza the town band is presenting a concert. We can see a little *nevería* (ice-cream parlor) in its center which has fresh green drinking coconuts piled on one counter. So as soon as we can wash the layers of dust from our faces and necks, we shall go down and each sample one. Then we will return to the hotel for a big dinner. We ate and ate and ate on the train coming down, but we now feel as though we had had nothing all day.

February 3, Manzanillo, Colima

George has been off the last three days making his studies along the lagoon and bay while I have found time to walk all over town. This doesn't take long to do, so I soon found myself down on the waterfront peering over a railing at the water. Being accustomed to the usual filthy water of ports, I was very much surprised to find the water here almost crystal clear. Every stone on the bottom at a depth of eight to ten feet can be seen perfectly.

Between them and the surface there are usually swimming a dozen kinds of small fish. Today some were velvety brown with an odd light eyespot placed topsy-turvily on their backs at the base of their tails. Others were pearl-gray with forked yellow tails. Still others were entirely striped in wide bands of alternate yellow and black. There were also broad black large-finned ones, and little needlefish with faces and snouts like knitting needles which usually swam in schools all lined up neatly side by side.

The only large fish I saw was about three feet long. It had a gar-like head and on the end of a thread-like tail, a foot-long waving whip structure. I had never seen such an odd whip end, and when I showed it to George, he hadn't either. We can't imagine what it is for. The body of the fish looks a little like that of an eel, but it is light blue-green in color with several rows of shining light dots. It swam slowly and deliberately back and forth along the bottom, sending the smaller fish nearby into panicky flight.

A hundred feet or so offshore a half-dozen brown boobies have been putting on their displays of precision diving. It seems good to see their brown-and-white patterns again. They are very common along the coast of

Peru. I have seen them there by the thousands when visiting their nesting grounds and photographing the nests and babies on the guano islands off the coast. They seem like old friends here, as it has now been several years since I have seen them.

When they dive for a fish, they stiffly hold their wings half spread at an acute angle, so that their bodies are arrow-shaped. With the terrific speed they attain, they dive entirely under the water. Then up they bob again and fly off to repeat the performance.

Our hotel dining room in Manzanillo is on the ground floor and is completely open to the street on two sides. The slightest breeze is therefore free to enter, and the diners and passersby can mutually regard each other over the potted ferns. We had an early dinner this evening before it was really dark. While dealing with the fish spines of a *mojarra frita* [fried fish; several species of the family Gerreidae exist in Mexican waters], I happened to look up and there sat a [House] sparrow on the back of a chair at the unoccupied next table.

From about three feet away he was eyeing us steadily, and especially the plate of bread before us. To see what would happen, I tossed a bit of roll onto the floor beside me. Señor Sparrow of the chair back didn't have a chance. As fast as the eye could follow, half a dozen of his brethren who were unobtrusively perched beneath the table on the lower chair rungs launched themselves on the tidbit. After a brief tussle it was divided three ways, and they all remained on the tiled floor looking hopefully at us for more.

We found we simply couldn't resist sparrows in the dining room with us. Especially when they are well-behaved sparrows, no one doing anything he or she shouldn't do. So we fed them on and off during the meal and are already looking forward to breakfast tomorrow so that we can invite them to join us again.

February 4, Manzanillo, Colima

After an early breakfast today we took the local bus line out to the Playa de Santiago, about twenty-five minutes' drive from town. George, fetchingly dressed in army jungle boots, bathing shorts, and a long-billed pilot's cap, then set off to study a freshwater lagoon not far away and to try and find out what the ducks on it are using for food. I walked around the little horseshoe sand beach and settled myself down at the far end.

From a high outcrop of rocks there I had a perfect view of the deep blue waters of the bay, the ever-changing colors of the mountains surrounding it, and the bird and some of the fish life on and in it. The atmosphere was one of peacefulness and tranquility, and less than half a dozen people were on the beach.

Two bare-legged, white-garbed natives fished for bait for several hours with dexterous flings of their wide circular *atarraya* nets. I could occasionally see the glittering silver *mojarras* as they slowly drew them in. Two brown boobies were also fishing, diving again and again deep beneath the water after small fish, and rising to the surface again with the bounce of corks. Dark blue crabs, speckled with rusty-red, scuttled and scratched constantly over the rocks around me. The smaller colorless transparent ones flung tiny clawfuls of sand here and there along the beach line of the receding tide, patiently clearing out their but recently filled holes.

Small fish darted everywhere in the clear water. Now and then they came up almost to the edge of the beach and swam between the legs of the few bathers, causing startled and unhappy outcries at their slippery touch. Although I saw no large fish today, it is said that even large sailfish periodically come well into the bay and show themselves above it as they leap and roll.

George came back well pleased. He had successfully collected not only the plants used as food by the waterfowl in the lagoon but also two kinds of tiny animals that they eat. One is an orange fairy shrimp about half an inch long. The other is a three-quarters of an inch yellow and gray slender fish with a rosebud mouth, which is usually projected and open as though in a state of eternal surprise. He found them both present in quantities and obtained good samples of each.

We had as an appetizer a tray of oysters on the half shell, just opened and served with limes and *salsa del Diablo*, "devil's sauce," which is as hot as it sounds. Then we had a very good lunch in the open-air dining room of the pleasantest-appearing hotel on the beach and returned to Manzanillo in the late afternoon.

We have just had time to prepare the specimens George collected, pack our bags, have box lunches made up to take the place of dinner, and catch the night train back to Guadalajara. It boasts a Pullman, American style, which is why we are taking it. But there is no diner, so passengers make their own provision for food beforehand, or go hungry.

February 5, Guadalajara, Jalisco

Our train, which was scheduled to arrive in Guadalajara at 7:25 A.M., actually pulled into the station somewhat after ten o'clock. We were fortunate enough to have had two sandwiches and several bananas left over from supper last night, so they served very well for breakfast. But almost no one else had brought along extra food, so most of the passengers had to go without breakfast until we arrived.

The muggy heat of the lowlands made it difficult to get to sleep in the airless berths last night. I was awake for several hours, during which time we passed through a number of small villages. I could see the tomato-red firelight from their little thatched houses briefly reflected against the walls of my berth, and hear a dog or two bark a short challenge at our rumbling passage.

In the very early morning I was awakened by a noise. I looked out to see the world muffled in a thick gray foggy wrap, and to hear disembodied voices talking somewhere not far away. But when I awakened again at seven o'clock, it was a beautifully clear crisp day with a cloudless sky. And we were moving slowly along the shore of a little lake.

Since our arrival in Guadalajara George has spent the rest of the day seeing people in our Consulate and talking with several local hunters here. I have been off shopping, which like for any woman is an occupation I usually thoroughly enjoy.

Guadalajara and several neighboring towns are famous for the handmade glassware and pottery which are produced there. So I decided to visit a glass factory near one of the markets today. I wanted to find more dark blue iced-tea glasses to match some we already have. Several rooms in the factory had shelf after shelf of glass dishes, each one differing slightly from its neighbors in both size and shape, because they are hand blown. Unfortunately, they had no iced-tea glasses to match ours.

I was asked if I wanted to go through the factory itself and see how the articles were made. In a long, open corridor, boys were cutting various widths of glass rods into measured lengths. Some of the rods were to end up as glasses, others as teacups, still others as pitchers, etc. In one large room two brightly burning furnaces were located so that several dozen workmen could ply their trade. Each man, I discovered, specialized in making just one kind of dish. Long lengths of tubular steel, each with one of the cut glass rods clamped in an end, were thrust into the furnaces. They were left until the glass became cherry-red with heat. At the proper moment, a helper took out a rod and brought it to one of the artisans. The maestro first placed a metal blowing tube at the free end of the glass. Then he twirled and pulled it out rapidly by the steel rod, and huffed and puffed at the blowing end until miraculously it had become a glass or vase or other desired object. When he tapped one rim with a small piece of metal, the lip of a pitcher would perhaps appear. Another bit of hot glass twisted against the other side added a decorative handle. All of it was done rapidly and almost silently, and row after row of the formed but still brittle dishes grew around each worker.

After they were all cooled, they were packed in an enormous kiln and baked at carefully regulated high temperatures for several hours. This made them less brittle and able to withstand ordinary use. The last step was the packing room where boys were making cases of peeled small branches, bound together with thin wire. In them, they packed the dishes, carefully cushioned with excelsior or crumpled newspapers, and they were ready to be shipped elsewhere in the republic. Before leaving, I bought a slender little decanter with six tiny liqueur glasses and a large round tray to match, all of a light blue-green color. I will have to supplement our iced-tea glasses at another time.

February 6, Pátzcuaro, Michoacán

We left Guadalajara this morning with the sky a brilliant crystalline blue. Row after row of high cirrus clouds marched across it, waving upward their feathery mares' tails. Against the light yellow-brown of most of the surrounding hills and countryside, the few irrigated valleys we crossed with their fields of chickpeas and wheat were startlingly blue-green in contrast.

Lake Chapala looked unusually muddy and murky today, and seemed to consist entirely of tints and shades of gamboge and ochre and mustard-yellows. As we drove high above it where the road has been dug and blasted out of the shoulders of a rocky mountain which rises steeply from the water, we could look down and see the shore in both directions for a long distance.

Along parts of it the fishermen have built little piers of stones, each about thirty feet apart and extending out perhaps ten feet into the lake. These divide the lakefront into little rectangular compartments, so that any one can be fished with a hand net without disturbing the small fish which may be in the others. The smallest fish found in the lake is the *charal.* It is a tiny fellow about two and a half inches long. At night when a fire is built nearby, it entices them shoreward and then they are caught in such pools with a circular *atarraya* throwing net.

In several villages today we passed odd, shining, silvery patches along the edges of the black-topped road. After we came to some that actually extended so far out toward its center that we had to swerve sharply to avoid running over them, my curiosity was aroused. We stopped for a moment so that I could see, and inspection proved it to be made up of several hundred of the little *charales.*

They were painstakingly laid out in rows on the dry, clean (comparatively) road surface to be dried in the sun. Although I don't care for them dried, they are very good when served fresh after being crisply fried and drained until they are free of grease. But at best they are something of a nuisance to prepare as they must always be beheaded and cleaned because they are so commonly parasitized by tapeworms.

We arrived in Jiquilpan in time for an early lunch at our favorite restaurant there. Then we drove on for miles along the edge of what is now a completely dry lake bed, but which in former times was an extensive bay of Lake Chapala. Here and there we went through little drowsing villages. Now they are stranded high and dry many miles from the water and have no industries that we could see, or visible means of self-support. They were, at one time, thriving fishing centers. I always wonder what a group of people, brought up to know but one trade, does to keep from literally starving when its entire environment and means of livelihood is, within a few years, taken away from it. This has happened over and over again in Mexico as many of its large lakes dried up.

The town of Jacona, a few miles from Zamora, was having a fiesta today. We had to force our way slowly through the crowds of people who spilled over from the sidewalks and completely covered the main highway. The old church and many of the houses were covered with yellow and white paper streamers. From the church, across a large enclosure to the road, there were fastened a dozen long rows of evergreen twigs. Just within the door of the church two priests, dressed in black and white robes, were sitting, one on either side of the door. Between them passed incessantly the entering and departing worshippers.

Little piles of fruits and vegetables and drinks were for sale along both sides of the road for several blocks, and everyone seemed to be eating something. We looked as we drove slowly along, but nowhere did we see any handwoven sarapes or clay pottery for sale, to us always the most interesting part of such fiestas.

We enjoyed it, as we always do, when we reached the area of lovely pine and oak forests with its many scattered volcanic cones all along the horizon in every direction. Soon Lake Pátzcuaro appeared before and below us, and we coasted down the last curving hilly slope to its level and to the little town of Quiroga. Quiroga was enjoying its weekly market day, so we stopped for a cold drink apiece and a brief look at the things for sale. At one little booth I found and bought a very well-carved mask in the form of a wolf's head, made from the soft white wood which is typical of the masks of the region. So now we have a new animal to add to our mask collection.

Turning right, we continued south around the lake to the posada at the village of Pátzcuaro, and it seemed good to see its hospitable patios and flower gardens and staff again. We took a leisurely walk through the gardens and then enjoyed an aperitif and a beautiful sunset on the second-story balcony overlooking the lake. Best of all, we had a most welcome and delicious dinner featuring the delectable *pescado blanco*.

Later in the evening a number of us settled in the *sala* (large room) before a blazing fire to enjoy its warmth in contrast to the sharp cold outside. However, before long a bellboy entered to tell us that there was to be a native dance performed on the terrace in front of the hotel.

The neighboring town of Tzintzuntzán is famous for its local dance, *Los Viejitos*, "The Little Old Men," which is always performed by all-male dancers. A group of seven of the villagers had come over to the hotel from Tzintzuntzán to dance it for the guests. George had seen it before during the fiesta there some years ago, but I hadn't. However, I had heard a number of times that it was one of the original dances performed by the Tarascans of the area when they needed rain, long before the arrival of the Spaniards. So we went out onto the terrace to see it.

The dancers varied greatly in age. Two were adults, one was in his late teens, one was about eight years old, and the youngest could not have been more than five. All were dressed in the white cotton pajama-type trousers and loose tops of the region. They wore straw hats over colored bandanas, which they had tied over their hair like baggy wigs, and leather sandals. Their masks were made to represent the faces of very old men and were brilliantly painted with hectic red flushes. Each mask had a tremendous beaked nose, wrinkles, and warts, and some lacked teeth entirely in spots while others had gaudy gold ones protruding.

Each dancer carried in front of him, and leaned heavily upon, an extremely curved wooden stick which he used as a cane. The five dancers lined up in order from tallest to shortest in a descending line. Then they started the dance to the time of a violin and drum played on the sidelines by their two companions.

The four oldest ones were very good, even the eight-year-old knowing his steps perfectly. The littlest one strove manfully, but sometimes found himself left a little behind, of course to the amusement of everyone. The steps were clever, with much wobbling of canes and shuffling of the dancers sidewise and forward with spry little hopping steps. Each member of the group had his turn at performing a solo in front of the line. Most of them simply repeated particularly intricate portions of the steps alone. When the tiny five-year-old very seriously and solemnly went through his solo, the last of the line and very well indeed, the applause of the audience was loud.

Then we all returned to the fireplace inside to thaw out for a few minutes after half an hour of shivering outdoors. Now George and I, after a brief bout of rummy in our room, are ready for bed since we must be up at an early hour tomorrow.

February 8, Pátzcuaro, Michoacán

We have spent two days on the lake in Chiquita while I made a collection of aquatic plants, which have been dried and pressed, and George did further waterfowl observations. And we both spent one day collecting desirable birds from the hills around the lake, and George has made them up into museum skins.

One thing we have been disappointed in is the fact that very few fishermen on the lake still use their strange butterfly-shaped fishing nets. Having heard and read for years about them, we were both looking forward to seeing and photographing them. But so far we have sighted only two boats, each with one of the nets, going home across the lake in the late afternoon. They still make their large dugout log canoes by hand though, and two are being constructed now along the edge of the canal by which we enter the lake in Chiquita.

We were interested this morning while out on the lake to hear and see at intervals bursts of gunfire. Tracing them, we found that a typical one came from a pair of natives paddling a dugout canoe. Two boats often worked in unison and slowly approached from two angles to drive a small group of coots or ducks together ahead of them. Then with an old-fashioned, muzzle-loading shotgun, a shot was fired into the birds. We observed that usually none was killed outright, and often that complete misses were made as the guns did not seem to be very accurate. At the most one or possibly two would be injured.

Then both boats gave chase, if there were cripples, with a man in each prow standing ready with a fifteen foot "throwing stick" for which the Indios here are noted. They threw these with considerable accuracy and after two or three tries often had their wounded coot or duck impaled upon it. Then, with a twist of the wrist its neck was wrung, and it joined the little pile of birds on the bottom of their canoe.

We talked with two men in one boat and examined the two throwing sticks which they had. Each one had three slender prongs, made apparently of filed nails or heavy wire and fastened with cord to one end of a bamboo shaft. On a bed of damp water plants in the bottom of the boat they had placed the three coots, one cinnamon teal, and one shoveler, which they had managed to get after four or five hours of effort. After a little bargaining we bought one of the spears for our collection of primitive weapons.[1]

I have been thrilled recently by my first view of [Northern] jacanas. They are amazingly colored birds which walk on outsized long toes on the aquatic vegetation where they feed. They have the odd habit each time they alight of holding their outspread wings together high overhead for a moment or two, which makes them resemble nothing so much as a bevy of ballet dancers.

The mature birds are a rich chestnut and have broad chartreuse centers and brown edges to their wings, and a bright yellow, shield-shaped structure on their foreheads. As they fly slowly along with extremely deliberate fluttery little strokes, the full beauty of their unusual wing pattern can be fully admired.

After some little effort, for they were quite wary, we collected two males to make into museum specimens. It is surprising to see how much smaller in the hand they really are than they appear to be when flying or walking about. An American who has lived in the region for many years calls them

The Tarascan Indians of the Lake Pátzcuaro region have long hunted waterfowl from hand-hewn dugout canoes using atlatls, which are locally called *fisgas*. These weapons consist of a hand-held "throwing board," with which a spear or dart is cast. This early photograph dated January 13, 1935, shows a Tarascan with an impaled American coot, or "mudhen." (Photo by Edward A. Goldman)

"Peter-walking-over-the water," which is a most appropriate name. They do seem from a slight distance to be doing just that. And since there is no visible means of support, one assumes they must have the invisible one of faith.

Pátzcuaro is fun to wander through in the early evening after work. It is built on hilly terrain, and old cobblestone roads still lead up hill and down dale with charming little views appearing before one at every corner. Under the enormous trees of its two plazas the public market informally spreads itself each Friday, and to a lesser degree on Sunday. When we are here studying on the lake and a Sunday comes along, we usually try sometime during the day to walk up to the plazas. Today we went, really planning just to look and probably buy nothing. However, the handmade earthenware pottery intrigued me, as it was identical to a piece I had bought previously. After some bargaining, I bought two pieces to complement the original. I could leave Pátzcuaro happy.

We have just finished packing Chico and plan to leave here early in the morning. If all goes well with his motor and other vital parts, we shall be in Mexico City tomorrow night.

February 9, Mexico City, D.F.

All did go well with Chico, and we arrived here today just at dark. We stopped for a picnic in the beautiful Mil Cumbres, "Thousand Peaks," region with its magnificent mountain scenery. And I had time there to make a collection of some of the lush mosses and lichens which I had yearned after on our way northwestward, at a time when there was no possibility of stopping for them.

Ever since I took a course in the identification of mosses while doing graduate work in botany at the University of Michigan, they have fascinated me. To the casual eye they look pretty much alike, all soft clumps of vivid green. But there are hundreds of differences among them really. Also, to be quite practical, they are the easiest group of plants to collect I know. No bulky plant presses or papers or blotters, and their daily changing for days afterward, are needed. At each collecting site all of the mosses found there are popped together into one small brown paper sack, and it is labeled as to place, altitude, date, etc. And that is all there is to it. There they quietly dry by themselves and need no attention from anyone.

We sighted nothing excitingly new in the way of birds or plants on the way back to the city, but had a pleasant trip nevertheless.

Many of the small houses we saw today and have seen on our various trips along the highways of Mexico, whether they are of the simplest mud-and-wattle construction, or more durable adobe or brick, are surrounded by small flower gardens. The people seem to love flowers, poor though their lives may otherwise be. Among almost any small group of huts or houses there will be at least one bougainvillea vine, brilliant in its red, cerise-purple, or more newly developed salmon-peach color. Geraniums grow in wild profusion whether stuck apparently in hit-or-miss fashion anywhere in the ground, or in more formal flower beds. Rosebushes do well where they are watered, and many of the more pretentious homes have large numbers of beautiful ones.

It is not at all unusual to see hanging orchids which have been brought in from the forests and fastened by cords to tree trunks, or put in little wire baskets that swing from the branches. In the garden of our hotel in Mexico City, which with its high altitude and arid climate is anything but a favorite spot for orchid growing, several of these plants hang from the enormous old trees and do very well. They delight the eye annually with their unexpected and lovely long stems of enduring flowers, which often last a full month before the blooms finally shrivel and die.

It is impossible to count, or even attempt to identify, all the varieties of wildflowers along the roads. Some of them are obviously the original ancestors from which our commonest garden flowers have been laboriously developed by experimenting horticulturists.

Midget marigolds form solid yellow and brown banks along the shoulders of the highway in many places. There are all kinds and colors of mints, fragrant and hung thickly with busily humming bees. Tiny yellow and pink cosmos, a minute zinnia, small single dahlias, beautiful coral thumb-size morning glories, and many, many more are sprinkled over the sides of the roads and nearby fields.

In February the orange trees are in bloom in many places and bear flowers and fruits at the same time. Their subtle and lovely scent is one of my favorites. We stopped and bought some of the fruit in a market today as we drove through a little village. They are arranged as they have been since time immemorial, in little piles of five, four below and one balanced as a pyramidal top. The native people, usually among themselves or in small out-of-the-way towns, still count by fives.

It is said that they do so because they started out long before the Conquest by using the fingers of one hand as an easy aid and unit of arithmetic, and in fact could count no further. Each group of five fruits or vegetables is counted as one. So five groups, each of five objects, total only five in their market reckoning. This can be definitely confusing until one realizes enough to stop trying to buy from them in dozens or half-dozens, after which it immediately becomes easy for everyone concerned.

CHAPTER 6

Off to the Mangrove Swamps of Guerrero

February 10, Mexico City, D.F.

WE HAVE gotten our mail from the American Embassy this morning and must spend a day or so answering necessary business letters and writing memos before we start out again. Also we must have some work done on Chico, and his motor checked again to see why in the world he is so slow.

Our next trip will be to survey the lagoons for waterfowl to the north and south of Acapulco on the Pacific coast. So we have to change our entire wardrobe from the heavy wool we needed for the cold of the highlands to our thinnest cottons for the humid heat of tropical beaches.

February 12, Mexico City, D.F.

I have been having a wonderful time today while poor George was slaving away at things which had to be done. First I went down to one of the oldest sections of the city near the Merced Market, where the many old houses and churches and people are always of interest. I did on-the-spot pen-and-ink sketches of a number of them for two or three hours. And then I returned to the garden of our hotel to make watercolors of several of the sketches.

It is always lots of fun and I enjoy it thoroughly, since my watercolors are done strictly for my own pleasure. If others like them, I am of course pleased. But that isn't a necessary part, by any means, of my continuing to take brush in hand.

After leaving the market, I made a beeline for Sanborn's restaurant, that haven of American cooking and drugstore delights for the homesick tourist. It is housed in an old Spanish colonial house which is one of the outstanding examples of the *mudéjar* (Moorish) style of architecture in the republic. It is thought to have been built in the year 1596 and is unique in having all of its exterior walls completely covered with the beautiful old blue-and-white handmade tiles for which the city of Puebla is famous.

It is always difficult in Sanborn's for me to choose among their ice-cream sundaes, their sodas, the wonderful fresh fruits, or something more exotic such as tamal pie. It is one of the few places in the country where we eat the ice cream and drink the milk, knowing with certainty that both are safely pasteurized. Our chances to take advantage of Sanborn's are all too far between, from my viewpoint, because we are for necessity away from Mexico City so much of the time working in the field.

Today I finally settled on a malted milk and fresh fruit salad for lunch, and they were delicious. I sometimes wish it were possible to carry along with us a portable electric icebox that would run on Chico's battery, and which I could fill exclusively with ice cream each time we leave the city. Although it seems that there is almost everything else welded or bolted somewhere about his body, that is one thing he unfortunately lacks. And I am sure, will continue to.

February 14, Taxco, Guerrero

We left Mexico City late this morning and drove out through the old and very large lava flow south of the city. Then we climbed the pine-covered mountain range between it and Cuernavaca, which is in the state of Morelos. As we topped the mountain, we could look down two thousand feet into the valley ahead and see Cuernavaca below us gleaming in the sun.

Every now and then as we descended the curving highway, we crossed the old handmade cobblestone road of Spanish colonial days. Many stretches of it are still, surprisingly, in very good condition. Over this old road the unhappy Emperor Maximilian and his Empress Carlota drove from their palace in Mexico City to Cuernavaca, where they spent much of their time in the 1860s during the brief time that France ruled Mexico.

After the winter chill of Mexico City on its high windy plain, Cuernavaca, twenty-seven hundred feet lower and protected by mountains from the north winds, always seems warm and mild by comparison. At the time of the Spanish invaders the village, then called Cuauhnahuac by its Tlahuica founders, was the capital of their nation. Cuernavaca, meaning "Cow's Horn," is a Spanish version of the sound of its original name. For once it is an appropriate one since it does spread over a horn-shaped ridge. The town was captured by Cortés in April 1521 and was included in the grant of thirty cities which Carlos V, king of Spain, later gave to him.

Two magnificent old plazas, filled with enormous and ancient shiny-leafed *Ficus* trees, share the honors of the center of the city. Along the side of one of the squares still stands the Governmental Palace, called the "Palacio de Cortés," which Cortés started to build in 1530 upon his return from Spain. The palace, a castellated affair, is one of the oldest buildings in America and is in a state of excellent preservation, although it is now unfortunately painted a bilious raspberry color. For years it was Cortés' favorite residence in Mexico.

It is also interesting because of a series of exceedingly powerful murals which the well-known Mexican painter Diego Rivera has executed on the walls of its loggia. They are colorful and strong, and in general depict various of Cortés' experiences in the country.

There are also several good hotels near the plazas, and we had a very satisfactory lunch at our favorite one. Then we continued along our way to Taxco.

On the dry slopes of Cuernavaca are the largest specimens I have yet seen of the "morning glory tree." This is a member of our familiar, climbing morning glory family of suburban trellises and vacant lots. However, it quite unexpectedly forms a thick woody trunk, which grows from twelve to fifteen feet high. Its typical flowers, lavender in color and exactly like those of the smaller vine, give way at once its relationship.

After some miles we crossed a wide valley and could see high mountains to the west of us, among which we knew that Taxco was perched. From here on, the road curves and twists upward, perfectly safe but continuously. The views behind us as we climbed were lovely, so we took turns driving Chico and both had a chance for ample sightseeing.

In the late afternoon we reached Taxco, which lies at 5,609 feet in altitude. We have put up at one of the many hotels on the mountainsides which look across and down on the city below. Now that the lights are on and the mountains stand out as rough black masses, it is a moving and beautiful sight from our balcony.

The old twisting, narrow, cobblestone streets are today just about as they were in the days of the Spanish colonial empire when its famous silver mines were producing such fabulous wealth for old Spain. It is no wonder that in late years the city attracted many artists. Literally every few yards a new view appears which simply calls out for brushes and paint.

I would love to settle down myself and spend a month here, doing watercolors. Unfortunately, however, George insists that our schedule doesn't permit even an entire day. It is too bad that just one of his important spots for waterfowl studies can't be conveniently located on the outskirts of Taxco, instead of, as they usually are, being miles away from anywhere.

The houses in town are tiled and are of various pastel colors in the Spanish style. They cling to the steep slopes of the hills in serried steps, so that from the sun-filled terraces of each layer one can look down over the roofs of the neighbors' houses, which lie just below on the slope. Everyone overlooks the center of the town's life, the plaza, with its towered church. The plaza is shaded with fine old *Ficus* trees, and benches have been placed beneath them to provide a pleasant place to sit and watch the town revolve around one.

Very fortunately a municipal law has been passed making all new buildings conform in architecture and type of material to the old ones. So it seems likely to remain for some time the gem that it is now. There are dozens of fine shops which carry locally fashioned objects in silver, copper, and tin, at reasonable prices. Shopping is really fun as there is such a wealth to pick from. I always find the final choices difficult though, because the designs are so original and the workmanship so fine.

February 15, Camp of the Magpie Jay, near Tierra Colorada, Guerrero

The drive down here today from Taxco has been definitely dull. The vegetation is more of the thorn shrub type, and everything is brown and bone-dry. The road is narrow, not as good as that from Mexico City to Taxco, and outside of one rocky gorge, there is little of interest to see. And it has been boiling hot all day!

The greatest sign of life or activity we saw along the entire road was a lone old turkey cock, or *guajolote*. With his face blue from passion and wattles shaking, he loudly gobbled his utmost defiance at Chico and the noise of Chico's unduly loud exhaust pipe.

We are now encamped about forty miles short of Acapulco, in a little valley off the road. A small pool of water at one side has made it possible for a dozen large trees to grow there, and so there is a good bird population. Innumerable trim little grasshoppers, smartly accoutered around the legs with scarlet trimming in contrast to their green bodies, hop and fly around us constantly with every step we take. We are planning to spend a couple of days here collecting birds from this type of habitat and region of the country for the museum.

Among the big shade trees in the coffee *fincas* (farms) on the Pacific coastal slopes of Guatemala, I made the acquaintance several years ago of a very large blue and gray jay with an unbelievably long pliable flowing tail, a long straggly black crest, and a terrific curiosity. I have never seen one since until today when three of these magpie jays [White-throated Magpie-Jays] flew, squawking, out to investigate us as we started to set up our camp this afternoon. So once again our camp has promptly named itself.

February 18, Camp of the Magpie Jay, near Tierra Colorada, Guerrero

We have both spent a busy and interesting three days here and have obtained a fine collection of typical birds. We use .410-gauge shotguns for this type of collecting with a regular .410 shell for distant shots of the larger birds. An auxiliary barrel, which holds a .32-caliber dust shot shell, slides in quickly for use when our target is a small or medium-sized bird and care must be taken not to injure its plumage by too forceful a charge.

There are various flycatchers here in large numbers, and we have collected seven different kinds of them now in this one spot. It has been a red-letter time for George, as white-winged doves are wintering in considerable numbers in the area. He has studied them for many years in the United States and in Mexico, both in the field and in captivity. So he is always anxious to learn any new facts that so far are unknown about their life history.

They are of considerable importance as a game bird in the southwestern part of the United States, where they nest and breed and pass the summers. But in recent years their numbers have been dwindling alarmingly, which is the reason he started to do research on them in the first place. Here in our camp he has been spending considerable time studying them to see whether the birds wintering in this part of the country are mostly members of the western or eastern races of white-winged doves, or both, and for other necessary data he wants about them.

Orioles are plentiful, especially the beautiful hooded oriole. And I have heard my first white-fronted [White-tipped] dove. It lives on a mountain slope across the road and periodically *whoo*'s in a low hollow tone. If I had not been told what it was, I would have thought it an owl. But once heard, it is not to be mistaken for anything else, so I shall know it from now on.

Living in our valley there are considerable numbers of a new bird. They are not large and have bold black head markings, and try though we both may, neither of us has been able to collect a single specimen. They don't seem to be especially wary. In fact, they sit upon branches in full view of us. The important point, however, is that they do so only when completely out of range of our number .32 shells. When the sound of a shot has once been heard, they all betake themselves at once to an acre of thick spiny brambles nearby, and there they stay.

We can then approach quite closely and watch them flitting about in its depths. Since it would take hours of machete wielding to find and retrieve one from there when shot, they are of course quite safe. And they seem to know it. We have played this game with them for three days now and are no nearer to winning than we were in the beginning. They have completely outmaneuvered us, which of course only makes us yearn for one more than ever.

There is a small shrub in flower around our camp which is apparently a favorite with all the hummingbirds in the neighborhood. Never have I seen so many before at one time. It is not unusual when we pass a clump north of our tent to see five or six of the little feathered mites, male and female and of several varieties, busily buzzing around each shrub.

It seems to each of us like taking unfair advantage of them, but it has enabled us to collect several very interesting ones. It is a most unusual and pleasant experience to be able to go there at almost any hour of the day and watch them in their feeding and dancing in flight. They chitter at each other in surprisingly loud, staccato notes when one approaches too closely a neighbor's luncheon flower for the latter's pleasure.

Hummingbirds have at least a dozen names in Spanish. All of them are highly descriptive of what they do, or how they sound. Common ones are *chupaflor,* "flower sucker"; *chupamiel,* "honey sucker"; *picaflor,* "flower sticker"; and *pájaro mosca,* "fly-bird." Their Náhuatl name is *huitzitzlin,* which sound like a flashing whir, itself.

The campesinos of the neighborhood have very politely gone about their business with little attention to us, and we have done the same. Last night we heard the sounds of music and laughter from a fiesta going on in the village some distance away, and for a while were afraid we might have unwelcome guests. Indios plus pulque or other intoxicants, plus foreign strangers, equal a volatile mixture that all too often spells trouble, and serious trouble. So we always try to be away from any small village on any known fiesta day.

February 19, Camp of the Boat-billed Heron near Playa de Cayaco, Guerrero

We broke camp near Tierra Colorada this morning and drove on into Acapulco to mail letters and stock up on some badly needed groceries. This accomplished, we started out north of town on the paved road that follows the seacoast. As far as the airport it hugs the cliffs, and one looks straight down into the beautiful blue bays, each fringed with milky smothers of surf. Then it curves inland so that the ocean, an adjoining sand ridge, and its inner lagoon are all from one to several miles away.

We continued along to the end of the pavement and beyond that on rough gravel, passing through a number of large coconut palm plantations. On the narrow road, surrounded on both sides and as far ahead and behind as I could see, were nothing for miles except the blue-green fronds of the palms.

It gave me the very eerie sensation of being underwater. The only other tree that has ever made me feel that same way is the olive. When walking or sitting in large groves of it in southern France, I felt the sensation even more strongly. I suppose it is because the waving branches overhead visually remind me by their motion and color of waves as one sees them when swimming underwater in the sea.

Since the lagoon birdlife is what George has come to study, we kept a sharp lookout for any side road that might take us in that direction. Just before we came to the thatched-and-wattled village of Cayaco, we could see the lagoon about two miles to our left, the nearest it had been for some time. So we asked in the village if there was a road to the playa or beach.

We were directed between several houses where there was not the slightest sign of a road, or the fact that any wheeled vehicle had ever passed that way before. Eventually, we came upon a very narrow track between endlessly fenced fields. We followed it slowly, expecting it every minute to become a few inches narrower or sandier so that we could go no farther. Finally, it deteriorated into a winding, machete-cleared cut in dense woods and thickets.

We reached its end and came out into a thirty- by thirty-foot clearing, hacked bodily out of a mangrove jungle on the very edge of the lagoon. We can look across this and see, half a mile away, the sandy barrier ridge upon the other side of which we know the sea is breaking.

All was intensely quiet as we arrived. We congratulated ourselves on the clearing being large enough to set up our tent and park Chico too, and proceeded to make camp. It seems already to be a very promising spot for new birds. George's excitement has been intense since within a few minutes of our arrival we heard the nesting song of a white-winged dove. It is for only the second time in Mexico this year, although we have listened for it in dozens of places. Soon several others swelled out their throbbing organ-like tones, and we know that a number of pairs are nesting or about to nest in the neighborhood.

I quickly unlimbered my recorder from my duffle bag. So by the time the birds stopped singing, at least one of the variations of the territorial song was safely down on paper for such posterity as may presumably be interested.

As darkness arrived, so did a medley of nocturnal birds and bird noises. Many of them are new to George as well as to me. The most interesting came from several unknown, and in the dark unidentifiable, birds which fished with loud splashes, and called and perched directly in front of our little camp. They were apparently completely unconcerned with the fact that humans were nearby.

George managed to collect a specimen, and it has turned out to be a boat-billed heron, and a new bird for me. So our camp is already named. This odd heron has a long, broad bill and enormous dark eyes that seem to be all pupil. Actually they may be largely that for the purpose of helping them see to feed in the dark. Its plumage is pearl-gray and black, and it has several untidy straggly feathers on the nape of its neck which it can raise like a crest when something excites it.

A fleet-winged [Common] pauraque swiftly zigzagged between me and

An adult white-winged dove tending two two-thirds grown squabs. Whitewings normally lay two eggs per clutch and often rear two young per nesting. Breeding activity usually occurs from May to August, and adults bring off two to three nests per season. Whitewings typically nest in highly concentrated colonies (sometimes exceeding a hundred nests per acre) in native thorn-brush. (George B. Sanders files, September 3, 1944)

the tent twice as I was washing the supper dishes. It startled me considerably, but added a new bird to my list. Just as we were about ready to crawl into our tent, well after dark, a new splashing arose in the shallow water a few feet in front of us. George turned the flashlight in that direction to see what was making the noise.

One lone, black-necked stilt was fishing for a belated supper and was obviously quite indignant at our unexpected interruption. He looked at us severely, uttered several short reedy cries, and calmly continued with his fishing. We watched him for a few minutes and then wished him good hunting and went to bed.

February 20, Camp of the Boat-billed Heron near Playa de Cayaco, Guerrero

This morning has been one I shall not forget in a long, long time. George awakened me when it was so early the mangrove thickets were only visible as blackish clumps against the lighter gray of the sky and water. Along the shore, not more than ten feet away from our tent, there were slowly stepping and feeding two large mangrove rails [Rufous-necked Wood-Rails], the first I have ever seen and our first birds of the day.

Lying in the sleeping bags on our stomachs, we watched for an hour or more until full daylight had arrived. We were treated to an assortment of silhouettes, flying, floating, and afoot, such as I had never imagined before. It was tropical waterfowl life at its untouched natural best.

The birds went about their accustomed activities of awakening, preening, and feeding as though they were entirely alone. As in fact they usually are in this remote place on the quiet lagoon. They went along as though our green tent and khaki Chico were nothing more than so much inanimate vegetation, grown suddenly perhaps, but nothing at which to be alarmed. There were mangrove and sora rails, boat-billed herons, stilts, grebes, egrets and herons of various types, assorted ducks and teal, one curious squirrel cuckoo, and even two roseate spoonbills flying so far out over the lagoon that I had only the briefest of glimpses of them.

After breakfast I went out on the lagoon by myself in Chiquita for about an hour while George was busy with other matters. I had some most rewarding sights as a result. I left the outboard motor ashore and used only a paddle so that I could slip along slowly and quietly along the edges of the reed patches and mangrove swamps that border the lagoon.

Everywhere waterbirds were going unconcernedly about the little businesses of their everyday lives. Snake-necked anhingas or water turkeys, in all stages of maturity and immaturity, were sunning and drying themselves on dead snags and tree trunks. Many of them held their wings akimbo by the hour, the better to get the sun. They didn't even change position as I glided along almost directly beneath them.

My biggest thrill, in the midst of many at seeing so many birds unknown to me and in such unfamiliar surroundings, was the sight of two very large black and orangy-salmon, zebra-striped waterbirds perched on a tree before me. They had extremely long rapier-like bills, which they occasionally opened and shut with a clattering noise. I had never seen anything like them.

I decided, therefore, to try and collect one as they certainly looked exotic, and to my hopeful mind, might as easily be rare too. After considerable effort I maneuvered Chiquita through the thick reeds far enough to be within the range of a .410 shell. I wasn't at all sure just what I wanted to do if the shot only wounded the bird. That long snapping bill looked as though it could easily penetrate a hand reaching to pick it up, and I had no other weapon with me in the canoe.

I shot, and one of the birds fell . . . out of my sight in the thick undergrowth. After what seemed an endless time, during which I imagined all sorts of ways in which it was escaping me, I managed to force the canoe first to within sight, and then within reach of it.

I was more than relieved to find a completely deceased specimen. Triumphantly I hurried back to camp with my booty and was somewhat crestfallen to find that George knows what it is. It is a Mexican tiger bittern [Bare-throated Tiger-Heron] and most appropriately named from its coloring and feather pattern. George says it will make a very desirable museum skin.

Later in the morning, and again in the afternoon, we put the outboard motor on Chiquita and cruised the length and breadth of the lagoon. George estimated the numbers and kinds of ducks there, and we took samples of the vegetation they were feeding on. However, ducks turned out to be but a small part of what we saw. There were Louisiana [Tri-colored] and little blue herons, white and wood ibises [Wood Storks], more tiger bitterns and anhingas, sora rails, coots, purple gallinules, black-necked stilts, snowy egrets, and assorted sandpipers, as well as blue-winged and green-winged and cinnamon teal, ringnecks, lesser scaup, shovelers, pintails, and ruddy ducks.

Back at camp a very large black, red-headed woodpecker [Imperial Woodpecker?] paraded up and down a nearby tree trunk while we were preparing bird and plant specimens. And a ruddy ground cuckoo [Lesser Ground-Cuckoo] came over and peered curiously at us for some time. Soon a lovely golden-brown hummingbird, such as neither of us has ever glimpsed before, came along and stopped to see what manner of flower my red shirt might be. A sweet-singing unknown tantalized us for over an hour by hopping and skulking though the thick mangrove tangles just out of our sight. Considerable ingenuity and teamwork finally secured him as a specimen, and he turned out to be, we think, some sort of chat [Gray-crowned Yellowthroat?].

It is now almost dark, and we are more than ready for our sleeping bags and a good night's sleep, as we have been on the go at top speed all day.

February 21, Acapulco, Guerrero

Today we are a pair of sadder but wiser people than we were when we went to bed last night. And instead of the good night's sleep we were planning on, we have never spent a worse one. It happened like this.

We were both asleep last night when at eleven-thirty o'clock I was awakened by the noise made by a group of Indios approaching the clearing. I quietly awakened George, and we settled down to see what was happening by the very bright light of the full moon. There were an old man dressed all in white, a younger man in dark clothes, two boys in their late teens, and a woman of indeterminate age. All four men carried long machetes, the knives which serve all purposes in Latin America from the felling of trees, to hoeing corn, to carving one's enemies.

It was immediately apparent that the younger man, whom we soon referred to between ourselves as "loud mouth," was overwhelmingly and vociferously drunk. His indiscriminate, ranting oratory came to a sudden focus as he realized that a tent and car were there.

"These people must be very rich to have all these things. Let us kill them, and then the things will all be ours," were the first coherent sentences we heard him say.

We both had loaded guns and extra shells with us in the tent, as is our habit when working in the field. But the last thing in the world we wanted to do was to use them.

Our tent is so made that only a straight front view is possible, and we were blind on both sides and the back. Any machete wielder could easily approach and hack through the thin nylon to the ground before we had any idea he was there, as bare feet are almost impossible to hear on sand. Also, even if with our first shots we could have disabled or killed in self-defense one or two of the group, any other fleet-footed member could have run back and told a wild tale to the village long before we could possibly have crept out along the narrow twisting lane in Chico. The only exit from where we were was through the center of the village, where a blockade could easily be set by enraged men who would not hesitate to kill us there. So we were pretty much trapped where we were.

And lastly, but most important of all, we knew that once we had fired a shot in such an affair, George would be persona non grata to the Mexican government and that his future work in Mexico would be at an end. And that is something neither of us wants to have happen.

So we waited unhappily to see what would develop next. Fortunately, we have always made it a practice to talk with and explain what we are doing to any inhabitants of a region we meet while working in their vicinity. Then their fears and imagination are at rest as to possible harm we might be planning to do their livestock or farms or people, which are the first things they usually think of.

They doubtlessly consider us at least moderately crazy to be so interested in birds and plants, but harmless. And the latter is the important thing from our point of view. This habit more than paid for itself last night, for the old man in white knew that we had been sent down by the United States government and what in general we were doing. So he attempted to dissuade "loud mouth" from his plan to murder us.

For over an hour they argued back and forth, sometimes almost on the point of physical violence themselves. We waited tensely in our tent, guns ready, to see what the outcome would be. And each moment we wondered whether the two teen-aged boys were moving silently up on the sides or back of us with their machetes.

Finally, "loud mouth" announced that he was going fishing, the original purpose for which they had come. Another hour passed while he ranted and raved in drunken style, and the two boys, under his orders, cut down young trees from the woods around us for poles and plaited a rope from split coconut palm leaves. Then he and the woman stripped and waded out into the lagoon, dragging the poles and the rope behind them.

For a long time there was little conversation. The boys and the old man kept a fire burning. We lay cramped and awake, wondering what the eventual outcome of the night would be. Then the two waders returned with a catch of fish, although our limited vision made it impossible for us to see how many or of what size they were. The fishermen were apparently cold and shivering, for they headed at once for the fire which was at one side of our tent about fifteen feet away, but out of our range of vision.

The pulque began again to circulate freely as the talk on everyone's part became loud and more boisterous. The two teenagers told interminable off-color stories which were greeted with much loud laughter. We were once again afraid of "loud mouth's" troublemaking under the stimulus of more alcohol. So we stayed awake and with every sense strained to the utmost until 4:30 A.M. At that time, after a further sharp argument between the two men about the advisability of killing us, they finally all rolled up in sarapes around the fire and one by one dropped off to sleep. About a quarter of five I did likewise, but George stayed awake and on guard all night.

At six o'clock when the first gray of dawn appeared, we crawled out of the tent with our guns before us and were up and facing them before they realized we had stirred. When they saw that only a man and a woman had been in the tent, instead of the several men "loud mouth" had probably assumed were there, the conflicting emotions on his face were something to see.

After a few sour looks and mutterings they picked up their things and departed. We spent the morning finishing up fieldwork on the lagoon. Then we broke camp and packed Chico, and shortly after lunch drove back out through the village to take the main highway without any further sign of trouble from its inhabitants.

We are now in a small hotel right on the sand of the beach north of Acapulco. We plan to continue our work and scout the first lagoon north of town and a lake to the south, with this as our headquarters. After last night's experience we are having to revise many of our plans.

Since any single, drunk native anywhere can cause us serious trouble at any time he comes upon us and he is so minded, we must rearrange things so that we no longer will be accessible for the trouble to occur. That means mainly no more camping and sleeping out in our tent in central and southern Mexico. We have decided that in most localities we can study only those places which can be satisfactorily reached within the daylight hours from a hotel, including the time necessary to drive there and back.

It will curtail things greatly and make completely impossible quite a number of places we had planned to study, because no hotels are nearby. But with one very close experience to profit by, we now know we have no choice. We are both alive, and we didn't have to shoot anyone. And that is the way we want things to continue.

Our little hotel here, although it has seen far better days, is arranged in just the way to suit us. All around the second story where our room is, there runs a fifteen-foot-wide open balcony. We can leave our work tables all set up on the balcony and work there, since we are the only guests. And it will be far cooler and give us much more working room than our small room does.

The sea-green fronds of coconut palms swish and sway with the sound of the patter of light rain, in clusters just outside the balcony railing. Beneath, scattered over the hard-packed sand, are the bamboo, one-room huts of a number of Indio families. The men are mostly fishermen, and there is almost always one sitting on a tiny stool under some palm weaving a new net with a hand-carved wooden shuttle. Tots of every age, and usually in an almost complete state of undress, toddle and play all day below us.

A little less than a hundred yards away the roaring boom of the surf echoes hour in and hour out, as the anything but placid Pacific Ocean throws itself against the steep beach. We have been out and walked along the playa late this afternoon. Aside from a few thatched huts, the hotel, and a restaurant, it is wild and beautiful.

Also, it has an attraction which I have seen before only in the Florida surf. Very commonly there are foot-long, slender fish hanging as though in a state of suspended animation at the top of the waves. But when they break in the next second, in frothy lacy patterns on the sands, they are gone. The fish never seem to be swimming violently to escape. They are just there one moment at the top of the waves, and the next, not. It has been a fascination for us that kept us watching this afternoon for an hour.

February 23, Acapulco, Guerrero

We spent all of yesterday catching up on some of our lost sleep and making museum skins of a number of birds which we had planned to prepare at the playa. Early this morning we put Chiquita in the southernmost end of the nearby Laguna de Coyuca, loaded our gear and ourselves aboard, and started off with the outboard purring smoothly. A dozen or so lesser scaup eyed us sleepily, some from as near as thirty feet away. They made no effort to move farther away. Foolish little scaup. Any local hunter could have picked them off with little more than a rock.

We headed north along the nearer shore, for the lagoon soon opens into a good-sized lake which is more than a mile across. Small thatched houses set on the seacoast ridge of sand in the shade of lofty coconut palm groves rim the shore and make a distinctly "south sea island" appearance. Here and there horses, cows, and pigs were standing knee-deep in the freshwater, cooling off from the already intense heat or having an early breakfast of aquatic vegetation. We shall never fail to be surprised to see the livestock in certain regions of Mexico commonly doing much of their grazing day after day on floating water plants, instead of on grass or ordinary pasturage. It is of course only where, because of one reason or another, there is no regular pasturage.

Slowly we continued to the end of the lake, stopping now and then to study the plants and waterfowl. Several birds on one old dilapidated pier turned out to be California gulls. And two killdeers took off from almost under my feet as we nosed Chiquita ashore on one point. Overhead the ceaseless flutterings and swoopings of hundreds of migrating barn swallows accompanied us hour after hour.

At the lake's end a narrow river leads still northward, so we continued to follow it. With every mile it becomes more beautiful. The vegetation is more and more lush, and as the stream winds through it, each curve shows another view of luxuriant tropical waving greens.

We stopped for lunch on the grassy bank of a small coconut grove where the river widens briefly until it makes a pool some two hundred feet across. Just as we were getting out cheese and peanut butter sandwiches and cookies, and wishing audibly that we had the climbing skill and machete necessary to get down two drinking coconuts, a native and his young son arrived on the scene as if in answer to Aladdin's rubbed lamp.

They were interested in what we were doing, so we told them. We added that we would certainly like a coconut or two if they knew how to obtain them. The response was immediate. Up the tall straight trunk of a tree scrambled the boy after a word from his father. He felt a number of them carefully and then threw down several which were just at the right stage for drinking. So we had delicious fresh coconut milk for lunch, and they both went on their way munching what may well have been their first "store bought" cookies.

The water looked so inviting and we were so terribly hot that afterward we went in swimming for half an hour or so. Then we started on up the river again, which becomes narrower and narrower the farther one goes. We passed several tiny villages of thatched huts where mothers were washing and children tumbling at the water's edge. Rounding one curve slowly and quietly, we came upon the first jacanas we have seen since those of Lake Pátzcuaro. True to their custom, they were tiptoeing along on the surfaces of floating plants.

We saw one giant kingfisher round-shoulderly hunched upon a small branch, peering down intently into the water beneath him. He was a ringed kingfisher, twice the size of our common belted one in the United States, with beautiful bright chestnut underparts except for a white throat. So I collected him for the museum as he flew across the stream in front of us.

He reminded me of what Mr. Alfred Newton, the English ornithologist, wrote about the kingfishers and the many medieval and classical beliefs that were held about them. Legend has it that in those days the kingfisher was considered to have been originally a plain, gray bird. It seems that it got its present bright colors by flying toward the sun on its liberation from Noah's Ark. Its upper surface at that time took on the color of the sky above it, and its under plumage was scorched by the heat of the setting sun.

Also, according to ancient fable, it had many virtues. The dried body would avert thunderbolts (museums should *never* be struck by lightning!). And of even greater value to the thrifty housewife, if put with clothes in a chest, it would keep them safe from moths. Or if hung by a thread from the ceiling of a room, it would point its bill to the quarter whence blew the wind. I offer this valuable information gratis for whatever good it may do any reader.

Finally we came to the dammed end of the little river and could go no farther without a portage. So we backed Chiquita around and saw everything all over again from the reverse direction.

It has been a very long day, as it was almost dark when we arrived again at the hotel this evening. Both of us, though fairly well tanned, seem to have unexpectedly acquired very red sunburned necks, knees, and noses. We have had a hearty steak dinner and are now ready to go to bed. And I know that I at least shall sleep soundly tonight.

February 25, Acapulco, Guerrero

We spent all day yesterday preparing the birds and water plants we collected along the river and lagoon the day before. Then we labeled and wrote up notes on all that we have taken in this vicinity, cleaned our guns, and did the dozens of other odd chores that always seem to be necessary.

Today we left Acapulco shortly after breakfast and drove east and then south to try and reach the Laguna de Tres Palos, the "Lagoon of Three Poles," where we know that many water fowl usually winter.[1] It proved to be an elusive lagoon to find. An American in Acapulco who swore that he had hunted there for years in season, mapped one route for us to the west of it, assuring us that we could drive to its very brink. After considerable time had been wasted driving back and forth, we discovered for ourselves that this road goes no nearer than a mile away from any point of the lagoon. And the intervening mile is either covered with the spiniest, most impenetrable thicket we have run into for a long time, or is soft, deep sand.

So, it being noon when this was finally settled, we then had to retrace our path around the northern end of the water and start down its east side. About this part of the lake we had been able to gather no information whatsoever. Several hours later, after progressing by fits and starts according to the directions given us by the last campesino asked, we drove triumphantly off the main dirt road for a quarter of a mile and through a small group of thatched houses, naked babies, and wallowing hogs to the water's edge.

By this time, however, the daily wind had picked up and was blowing so hard from the north, and such large waves were running across the lagoon and breaking upon the shore in front of us, that we knew we could never satisfactorily cruise it in Chiquita. So George did all the observations he could from various places which we could reach along the shore.

Then we drove back along the main road in the direction of Acapulco to a likely spot to watch the late afternoon return of the land birds to their evening roosts, and to listen for possible singing white-winged doves. Flock after flock of squawking parrots and parakeets flew over us, some perching in nearby wild fruit trees for a noisy snack before turning in for the night. I collected one medium-sized parrot I haven't seen before, which has beautiful dark blue wing patches instead of the usual red or yellow ones.

We drove down the last mountain slope and into view of the unbelievably beautiful harbor and natural setting of Acapulco just as the setting sun in front of us was trailing sulphur and violet clouds across the sky against a cataclysmic crimson background. It was one of the most striking sunsets I have ever seen.

February 26, Taxco, Guerrero

We drove as far as Taxco today on our return to Mexico City. As before, it was a hot uninteresting drive through arid, barren country. We stopped for a sandwich lunch just off the road in a spot that looked as though no humans or lower animals had ever set foot there for lack of water or decent soil. Yet within five minutes we were surrounded by a cloud of biting little black flies.

They didn't bother George at all, rather unfortunately I thought. For my legs and arms were soon simply peppered with tiny raised blood blisters, each one the site of a bite. From previous experience with the little monsters in the *fincas* of Guatemala, where they are called "coffee flies," I know that each bite will itch for days if I don't scratch it. And it will rise into a welt and itch for about two weeks if I do. Eventually I always do.

We have had a fine time this evening. For some years, in fact ever since a friend returned from a vacation trip to Mexico with one, I have had a great yearning for a silver pin made in the shape of an owl with two large round amethysts set for its eyes. And I have never been able to find one anywhere.

Tonight I decided to search for one. We have been in and out of all the fascinating shops, large and small, where the hand-wrought silver articles for which Taxco is so famous are designed and made. It was lots of fun, as many of the articles are original and lovely. However, it was not until we were in almost the last unvisited shop on the last street that I found my owl. It is a beauty, and I am so pleased with it. Never has a real *tecolote* looked out upon the world with deeper, more fathomless eyes.

We had a very good dinner in one of the old Spanish colonial homes which has been remodeled as a hotel. The diners sit at small tables around the edge of a flower-filled patio in which a small fountain splashes merrily in the center. Then we went back up through the narrow winding streets to the main plaza. With its benches and mammoth green *laureles* trees, it is a very pleasant place to sit and watch the life of the town for an hour or so before going to bed.

Through the lighted windows of a restaurant on one corner of the plaza we could see a splendid collection of the wooden masks used in the native dances of various regions in Mexico. For a number of years I have been collecting handmade masks worn in Latin American countries.

I particularly like those of tigers and deer, which are widely used in dances in many countries. Although we have managed to find and buy several very different tiger masks in Mexico, we have not yet been able to obtain anywhere in the country one of a deer. So I was envious to see a deer mask in this collection, but unfortunately out of our reach. Someday I hope Fate will be kind enough to put an available one in my path to fill the empty place that is waiting for it along our long hallway where the others are hanging.

February 27, Mexico City, D.F.

We had an uneventful return trip from Taxco, marked only by a very fine lunch at Cuernavaca and a visit to the old Borde estate there. This is an old house with extensive gardens built by the Frenchman Joseph le Borde of Taxco fame, who came to Mexico in 1716. It is now open to visitors for a small fee and gives some slight idea of the sumptuousness of the wealthier families' former lives.

Even today the old trees and walks and ponds are enchanting and well-planned. There are several pools, a five-hundred-foot lake, cascades, fountains, bathing pavilions, arbors, terraces, little summer houses, and a chapel of its own. One corridor leads to the rooms that Emperor Maximilian and Empress Carlota always occupied during their visits to Cuernavaca.

In one place the cobbled roadway around the grounds is raised so that beneath it there is a series of small rooms, each one with a grilled window. Perhaps it was a useful method of providing unobtrusive housing for some of the servants. It is certainly reminiscent of and similar to the ingeniousness of many of Thomas Jefferson's arrangements at Monticello in Charlottesville, Virginia.

The gardens, though not now well cared for, are still pleasant to walk through. One stately old giant of a tree was in flower. The grass below it was covered deeply with its five-inch-long feather-duster blooms of a beautiful rich rose color. It is a tree I have never seen before, and I am anxious to find out what it is.

We are back in our favorite hotel in Mexico City once again and will stay here a few days. George must bring his office work up to date, and we must repack and load the equipment we will need on our next trip to the lagoons near Veracruz on the Gulf of Mexico. Then we will have some very necessary things done to Chico. We still think that something is definitely wrong with his motor and that he should have far more power and speed than he does.

George Saunders, binoculars in hand, standing alongside their jeep, Chico, with their canoe, Chiquita, strapped firmly on top. (George B. Saunders files)

Snow geese were among the many species of waterfowl that George Saunders was instructed to inventory during the 1948–49 trip. The species is among the northernmost

breeding and southernmost wintering North American geese. Because they were often difficult to approach on foot, their numbers were best estimated from aircraft. (Photo by Roy E. Tomlinson)

Dorothy seldom missed an opportunity to visit the *mercado,* or market, whenever their schedule allowed. The *mercado,* a collection of centralized shops and stalls where food and nearly every other household necessity was sold, was along with the local church the center of a town's economic and social activity. In this photograph of a typical *mercado,* note the densely packed outdoor stalls, the absence of motor vehicles, barefoot or sandal-clad shoppers, woven shopping baskets, and the colorful clothing. (George B. Saunders files)

The burro was still a common means of transportation during the Saunderses' day. This seemingly contented sturdy beast is outfitted with crudely made but serviceable saddle, cinch, stirrups, and bridle and is tethered to a palm tree before a native dwelling. (Photo by Roy E. Tomlinson)

Dorothy thought the vermilion flycatcher was one of the most beautiful birds that they saw during their trip. This species is plentiful around streams and small ponds throughout most of Mexico. (Photo by Roy E. Tomlinson)

Pyramids of the Sun and Moon at the site of Teotihuacán thirty miles northeast of Mexico City. At its height of development between A.D. 300 and 700, the city had some 100,000 to 200,000 inhabitants and is regarded as the first true metropolis of Middle America. The larger of the two pyramids, that named for the Sun, measures seven hundred feet square

and is as high as a twenty-story building. It is the largest human-made structure of the Americas dating to the pre-Columbian era. Atop both pyramids were platforms upon which human sacrifices were offered to propitiate the gods. (Photo by Roy E. Tomlinson)

This apparently abandoned simple one-room dwelling with a dirt floor typifies the residences occupied by many rural folk during the 1940s. (Photo by Roy E. Tomlinson)

Chico at a typical campsite. Note that he is fitted with a front bumper winch to assist in extracting him when high-centered or stuck in holes, mud, and dry or deep watercourses. The nylon military surplus tents are of World War II vintage. (George B. Saunders files)

Dorothy with eight ducks collected for scientific specimens. Presumably she shot these herself, as she was a very proficient marksman with small weapons. In her bag are a certain canvasback drake, a pintail drake, a probable gadwall, and possibly two ruddy ducks. The rather large duck may be a Mexican duck (formerly regarded as a full species, now regarded as a form of the mallard). (George B. Saunders files)

A typical scene along the west coast of Mexico. The Saunderses were awed by the scenery

during their trip. (Photo by Roy E. Tomlinson)

Lake Pátzcuaro, Michoacán, viewed from the south. The beautifully situated lake has been one of Mexico's most important waterfowl wintering areas, although in recent years the numbers of birds have diminished. At the time of the Saunderses' visit, Tarascan Indians still hunted waterbirds on it with atlatls, an ancient weapon that elsewhere was usually replaced by the bow and arrow. (George B. Saunders files)

The Saunderses encountered colorful jacanas in many marshes of central and southern Mexico. Dorothy was particularly delighted whenever she saw one. This one is treading lightly over a mat of water hyacinth interspersed with stalks of horsetails. (Photo by Roy E. Tomlinson)

Popocatépetl (17,887 feet; left) and Iztaccíhuatl (17,338 feet; right), dormant snow-capped, sometimes still-active volcanoes southeast of Mexico City, punctuate the horizon. The silhouette of the latter is often compared to the outline of a reclining woman, whose head is to the right. The neighboring peaks are, respectively, the seventh and eighth highest in North America. (George B. Saunders files)

Oxen were still used as draft animals for pulling wheeled wagons and plows during the late 1940s. The Aztecs and related native peoples lacked beasts of burden as well as the wheel. (Photo by Roy E. Tomlinson)

Dorothy Saunders feeding greens to a pet white-tailed deer named Bambi that was kept by owners of a lavish tourist hotel in Fortín, Oaxaca. After several days in the field the Saunderses checked into hotels for a day or two to refresh themselves with baths or showers, softer beds, and fine food. (George B. Saunders files)

The sleek phainopepla, a favorite of Dorothy's, favors mesquite brushlands and feeds extensively on mistletoe and insects. (Photo by Roy E. Tomlinson)

George is perhaps best remembered, ornithologically, for his detailed studies of two species, the white-winged dove and the meadowlark. During their 1948–49 Mexican trip George continued to gather information on both, while Dorothy musically recorded for the first time the calls of the whitewing. These whitewings are perched atop a mesquite tree. (Photo by Roy E. Tomlinson)

George tediously preparing bird study skins for the National Museum's reference collection. The painstaking work required much skill and experience, along with taking measurements and keeping records for each. (George B. Saunders files)

Chachalacas are gregarious inhabitants of dense brush. Their name is onomatopoetically derived from their raucous calls, which awakened the Saunderses many mornings when camped out. (Photo by Roy E. Tomlinson)

Yuccas are integral to the vegetational makeup of Mexico. Fibers from yucca leaves have been used for centuries by Indians for clothing, rope, and household goods. These were photographed in the state of Tamaulipas. (Photo by David Dolton)

CHAPTER 7

To the Gulf of Mexico

March 5, Córdoba, Veracruz

WE LEFT Mexico City early today for Veracruz via the cities of Puebla, Orizaba, and Córdoba. The drive from Mexico City to Veracruz is an awe-inspiring one, much of it being along the oldest Spanish road in the republic.

The first part of the highway from Mexico City meanders across the plain and is edged with tremendous old cedar and eucalyptus trees. Then the mountains are reached, and the road winds back and forth up and over them at ten thousand feet with keen mountain air, lovely little high meadows, and large pine and oak groves for constant accompaniments. We chugged down the mountains, Chico proving to be not a bit faster than before in spite of a new carburetor and a thoroughly checked engine.

Sometime before we reached Puebla de Los Angeles, the "City of the Angels" which was founded by the Spanish in 1532, we knew that we were nearby, as we saw dozens of lovely old Spanish colonial churches. Many are still complete with rows of the beautiful colored handmade tiles for which Puebla has been famous for almost four centuries. The churches are dotted here and there over the valley and its nearby hills like so many oversized bouquets of flowers. Each has at least one rounded, brightly tiled dome and a tall graceful bell tower. We were soon never without several in sight, many of them real architectural gems.

The day was brilliantly clear, and we had a superb view of the two great snow-capped volcanoes that hang above Cholula and Puebla. Our road wound first not far (apparently in the crystalline air) from the base of Iztaccíhuatl, which means in Nahuatl, "Sleeping Woman." It rises 16,200 feet above sea level and literally appears to be a reclining snow-covered woman. There are three peaks which form her head, breast, and feet. Lower saddles fill in the body proper, making clear her femininity. We could easily understand how the Indios came up with her name centuries ago.

Within a very few minutes our road curved sufficiently to show us her companion, snow-topped, cone-shaped Popocatépetl, peering over from behind Iztaccíhuatl. "Popo," as it is familiarly called, is the Nahuatl for "Mountain of Smoke." And today we saw clearly a thin slight veil of smoke rising steadily from its southern lip.

It has been intermittently active through the years, with its last outburst in 1921. Cortés first discovered the hundreds of thousands of tons of excellent sulphur in it and used part of them successfully in the manufacture of gunpowder to continue his battles against its human neighbors.

Soon afterward I added a new bird to my life list when a black-headed grosbeak, loudly marked with dashing white patches on black wings, a black head and golden-brown breast, flashed across the road before us and alighted on a nearby shrub.

Unfortunately we had no time to stop over in Puebla on this trip. We continued directly through it and along a valley which became more and more arid until we reached Tehuacán, the site of famous mineral springs and bottled water. Then the road curves sharply back upon itself and up and over a ridge eastward to the great Mexican Plateau. This is covered with cacti and not very interesting, and we were both feeling rather somnambulant as we continued along it.

Suddenly we were not only both wide awake but considerably startled besides. Our road without warning came directly to the brink of the plateau, and all at once there was nothing in front of us but air. Two thousand feet below us we could see several miniature towns in the valley. Never have I, in spite of having driven over many of the steepest Andean roads in Ecuador, Peru, and Chile, seen a piece of engineering to equal that road.

In many ways the view reminded me of the Central Highway from Callao up the Andes to Orroya in Peru. High mountain ranges sweep ahead on both sides and fade away into ever lighter misty outlines. Everything on the mountain and down their sides to the long narrow valley bottom is the dun-beige color that a lack of water brings to even the most fertile soils. But in the valley where a stream rushes, patches of clear emerald-green fill in almost every available meter. The line between where water is and isn't available could very nearly be marked by the cut of a knife blade.

Down the escarpment we wound like a whirligig in low gear and with the brakes on a good part of the time besides. Around one hairpin turn after another we went, on a well-banked, perfectly safe road as long as one's brakes, gears, and various wheels all continued to function. It took us over twenty minutes to reach a point at the bottom of the drop almost directly beneath the spot we had started from. And the little toy villages we had seen from above turned out to be respectably sized towns, and even included the city of Orizaba.

It was an experience well worth having. But we knew right then that with Chico's underpowered engine and our load, we hadn't a chance of ever making it back up the plateau by that same road.

The city of Orizaba, at an altitude of 4,028 feet, is in the temperate zone halfway between the *tierra fría* [cold region] of the Plateau and the *tierra caliente* [hot region] of the coast. It is an important manufacturing center and is also popular as a resort with both the frost-nipped inhabitants above it and the heat-wilted ones below. Its heavy rainfall brings out many beautiful flowers, as is also the case on down the descending valley at Fortín and Córdoba. We were soon surrounded by coffee, bananas, oranges, pineapples, and many kinds of flowers, including, unexpectedly, large fields of azalea bushes in pink and white bloom.

Fortín is the well-publicized site of a lavishly outfitted tourist hotel, noted principally for its swimming pool, which is covered daily with a floating layer of gardenias for customers to swim among. American tourists in large numbers with all their too frequent uninhibited antics once they have left the borders of the United States behind often set our teeth on edge. So we went on five miles to spend the night in little Córdoba. Now we are glad we did.

It is a pleasing, entirely restful little town which was founded in 1618 by order of the Spanish viceroy Don Diego Fernández de Córdoba. At about twenty-seven hundred feet in altitude, the climate is a balmy one which has already relaxed us into a shoestring-like state. We have found, facing the main plaza, a nice hotel which was originally built in 1687 as a private home. A plaque in one wall says that in it the Treaty of Independence of Mexico was signed on August 24, 1821, by General Augustín de Iturbide and the last Spanish viceroy, Don Juan de O'Donojú.

It has beautiful old tiles and is pleasantly arranged around a patio with a tinkling fountain in the center. On the front of the second story is a wide balcony with chairs and tables where we sat before dinnertime and watched the life of the little plaza move at a sedate pace below us.

The Palace of Government is diagonally across from us on another side of the square. We were highly amused to see a single soldier solemnly marching up and down in front of its doors, rifle over shoulder, and snappily about-facing at each end of his beat. It is the first time we have seen such martial efficiency in any town or city of the republic, aside from Mexico City. Usually it has been our observations that rifles are used more as props to hold up the sagging bodies of whatever thoroughly uninterested-looking members of the armed forces there may be about.

But not in Córdoba. Soon the quiet air was split by the brassy notes of two buglers. A lineup of uniformed men took its place in front of the palace doors. Unfortunately the big pillars kept us from seeing the center area, but apparently someone made them a speech while they stood at attention. The bugles sounded again and everyone disbanded, except the lone sentry who continued his brisk march.

The owners of the hotel have a year-old white-tailed fawn. The fawn is inevitably named Bambi, and is given the run of the downstairs. He is extremely tame, comes by name, and allows himself to be petted. We of course are enchanted with him, especially so when he very companionably curls up at our feet when we sit in the downstairs patio.

Since there was time before dark, we went to the nearby market to see what they had on display of interest. There was considerable, including pile after pile of juicy-looking fresh fruits and vegetables. We were fortunate enough to find one woman from the nearby Indio village of Amatlán, a small town where their own laws, customs, and language have been retained since before the days of the Conquest. She wore their typical loose-flowing robe of white cotton decorated with dark blue braid, for which we had been told to look.

On our way back we stopped for half an hour to watch the magnificent mountain, Pico de Orizaba. It is a perfect snow-capped volcano of the shape of Japan's Fujiyama. It was called by the Aztecs Citlaltépetl, "Mountain of the Star." And it is the place where in their belief the body of Quetzalcoátl, "Feathered Serpent God," was brought after death and consumed by divine fire, his spirit going heavenward in the form of a peacock.

Since the Aztecs believed that his spirit would someday return to Mexico, the mountain is still venerated in a religious sense by many of their descendants. Our view of it in the early evening light will always remain in my memory as one of the most spectacular sights I have ever seen. It is, at 18,225 feet, next to the highest peak on the North American continent. And until Mount McKinley [Denali] in Alaska was measured, it was long considered to be the highest.

The wittily acute and widely traveled lady, Madame Calderón de la Barca, wrote with her usual penetration one of the aptest comments on it I have ever read. She said, "The mountain has what mortals rarely possess united, a warm heart, with a clear cold head."

In disembodied majesty it rose behind the town this evening, and no other word describes it so nearly as *ethereal*. Although the skies were clear behind it, its base was covered by a distant broad band of light haze so that it seemed literally to rise without foundation from the air. Lovely lights constantly changed upon it and in the sky behind it. We stood, each alone with our thoughts and without speaking, for a long, long time.

When the sky was finally dark and the mountain only a black mass against it, we went on to the hotel, bearing as a gift for Bambi a head of lettuce from the market. Bambi was highly pleased and daintily ate it at once from our hands as a luscious change from his daily dry hay. He then immediately lay down by one of our chairs and began to chew his cud, and I learned something I hadn't known before. A deer chews with delicate but steady little sidewise motions of the lower jaw while the upper one oddly enough remains completely stationary.

We shortly thereafter decided to have dinner in the hotel's restaurant, and a fine one it was too. This long room looks out through grated windows onto the plaza so that diners may see and be seen by the strollers there. They very fortunately had one of our favored dishes, *sopa de ajo*, on their a la carte menu, so we both started out with it.

I must confess that the first time I read "garlic soup" on a Spanish menu, I imagined that it would taste and smell perfectly horrible. But, since trying new dishes is one of my pleasures, I figuratively steeled myself and ordered it.

And now there are few soups that I like so well. Its flavor is very delicate, and the odor, once it is off the stove, is nonexistent. Its beauty also lies in the fact that it is extremely easy to prepare and takes but a little time to do.

All during our dinner the air was filled with the delicate tropical scent of unseen night-blooming flowers, one of the loveliest odors I know. After a final goodnight look at the mighty Pico, we are going to bed early tonight, as we are emotionally tired from all the magnificent scenery we have seen today.

March 6, Veracruz, Veracruz

We left Córdoba this morning and soon were immersed in even more tropical growth than what we saw yesterday in the valley above it. We drove through miles and miles of plantations of sugarcane, coffee, tobacco, mangoes, bananas, and papayas, which crowded up to either side of the road.

The coffee was just coming into white-rowed flower. Its sweetly insistent scent, almost equal in my opinion to the lovely fragrance of orange blossoms, insinuated itself into our senses from time to time. The stalks of the oddly upside-down banana flowers had, as always, a full quota of buzzing hummingbirds and bees competing for their dripping nectar.

Soon we saw coming along the side of the road toward us a little brown burro which was followed by an Indio on foot. On each side of the little animal an empty wooden fruit crate was tied with rope to serve as pannier. I was just thinking idly how odd it was that the man should be walking instead of riding, when I saw above the rim of one of the boxes a tiny much-worn straw hat and a pair of enormous dark eyes. His little son, aged perhaps two years, was having the ride, attired in nothing but a wide-brimmed sombrero and a look of intense interest in the world. As far as we could see them in the rear-view mirror, burrito clip-clopped steadily along and Papa trudged behind.

The country became more and more rolling, and extremely arid with a dry-woodland vegetation. Shortly before we crossed the railroad from Veracruz to Alvarado, we came upon what at one time was a very large pre-Conquest settlement. On both sides of the road as far as we could see, and scattered along it for several miles, were the large truncated pyramids of their former temples and other buildings. Grass and tree-covered they now stand, alone in the middle of fields, pastures, and woods, apparently untouched since the time they were abandoned.

How much we wished we had the time to investigate a few of the mounds, or at least to map the entire site and find out something more about it and its inhabitants. One incongruous note was struck near the end of the ruins when we found, close to the road, that someone had used a fine old pyramid for the base of his home and had built a completely modern stuccoed house on its top. I couldn't help wondering if the bones of its former highly revered inhabitants weren't whirling unhappily beneath at the sacrilege.

Not far beyond this we saw as sweet a sight as the recent house-on-a-mound wasn't. The sun had become very hot by now, and both the temperature and humidity were high. In a small pasture which had no tree or other shade of any kind, an obviously brand-new baby lamb was lying. Its mother apparently realized that the sun, hour after hour, was too much for it. So she had carefully stepped over, parallel to and above it, so that as it squatted on its haunches like a little dog beneath her, she completely shaded it. There they stayed motionless, two curly-haired, carbon-black figures, with Mother taking all the very hot sun on herself to spare her baby.

We crossed the river at Boca del Río. Along one bank sprawls a quaint little town of small thatched houses, coconut palms, and open-air eating establishments which are noted for their fine seafoods, around a minute plaza.

Then our road curved down and ran directly along the edge of the sandy Gulf beach. Three groups of fishermen have built themselves three large single-roomed community thatched houses where all live together, men, women, and children. Outside, their long slim boats were pulled up on the beach, and row after row of café au lait–colored handwoven nets were strung on sticks to dry.

Just as we came along, they were hauling in the last hundred feet of a tremendous net which they must have spread some hours before. The men were solemnly marching in a row through the shallow water and along the sand to their drying poles. With the long, rolled net looped over the shoulder of each man, they looked like a grotesque parody of Vassar's famous daisy chain.

Three pelicans floated fearlessly on the surface of the water within the net's borders and frequently reached down a long broad bill to snap up some small fry, caught by and rightfully belonging to less lazy folk than they.

We stopped for a moment to see the catch. There were several hundred fish, many of them mackerel, pompano, and redfish. Nearby they had somehow managed to land a real prize, a large sea turtle, turned over on its back.

We have decided to stop at a large deluxe resort hotel on the beach several miles short of the city of Veracruz proper. George's studies are to be on the lagoon between Veracruz and Alvarado, forty miles to the south. And the road we must take to get there branches off from our main Córdoba highway about eight miles back along the route we came today. So by staying here, we won't have the traffic of the city to slow us up twice a day as we leave and return, and we will save about ten miles of driving every day.

It is an elaborate hotel, built upon terraces of various heights so that it resembles nothing so much from the sea side as a multilayered, heavily frosted white sugar cake. But the rooms are pleasant and have a fine breeze, which is very welcome in the intense heat.

The wide porch outside our windows will make a splendid place for George to prepare and study his birds, and for me to press and dry the lagoon and marsh plants which we collect. So we will settle down with the Mesdames Vanastorbilts and their consorts for a few days. Although our spare time to view them and theirs will be very much limited.

March 9, Veracruz, Veracruz

We have spent the past three days on the Laguna de Alvarado studying the ecology of its duck foods and collecting specimens of them. We drove down the forty miles to Alvarado through extensive pineapple plantations and palm groves and, unexpectedly, through one good-sized area of natural shortgrass savanna.

Alvarado, which we had expected to be a rather attractive coconut palm–encircled little fishing village, is distinctly not. We found it quite unlovely. It is hot and dirty, which of course by itself is not important if anything of interest is there. But there is nothing in the least picturesque about it. It, as well as its inhabitants, seems to be entirely lacking in either friendship or charm. And there are numbers of persistent little boys who don't understand the meaning of the word *no*, in any language.

We finally found a little place back of a building in the town where we could park Chico and put Chiquita in the water within carrying distance. So the first day we surveyed the lower part of the lagoon near its entrance into the Gulf and were happy to leave Alvarado sweltering in its own juices by late afternoon.

The second day we put Chiquita in the water about halfway along the length of the lagoon and spent that day studying the vegetation of the bay in that part. About noon we came upon a heron, out a short distance from shore, standing on a mat of floating water weeds. He stood stolidly, surveying the world and us with an air of complete boredom as his half-sinking raft slowly took him along with the current.

Today we have finished the lagoon end nearest Veracruz, gaining access to the water via the yard of a very pleasant native family who lives in a little thatched hut on the water's edge there. The water of the lagoon is very brackish, but there is a great deal of duck food in it.

We saw one raft of several hundred baldpates, although most of the ducks had already started their northward migration. A few coots leisurely and fearlessly moved back and forth in front of us after the manner of their kind. Great blue herons and American and snowy egrets stood stiff-legged in the shallows, looking at us with indifferent eyes as we paddled past. Cicadas, unmistakably sopranos, carried on their tunes around us from the shore for interminable intervals, all at fortissimo pitch. Two little golden-orange butterflies fluttered out over the water and flew directly over our bow, spinning together like a pair of ballerinas.

As we were driving back to Veracruz late yesterday afternoon, we passed two little burros just outside a fence by the road. They were simultaneously scratching each other's necks with their bared teeth, in careful little nibbles. Never have I seen expressions of more supreme contentment on any living faces.

When we reached the fishermen's communal houses, they were again pulling in their nets. The [Brown] pelicans were calmly exacting their daily toll of the small fish. Overhead an osprey was circling, although he already had one large fish held horizontally beneath him, torpedo fashion, in his sharp talons. And he couldn't possibly have managed another one too.

We now have a fine lot of specimens of plants drying on the hotel terrace, and several mollusks which the ducks apparently feed on too. The more resplendent guests look at them, and us, rather quizzically as they go past, but so far have been too reserved to question us about them.

After a very fine dinner this evening, for while the service leaves something to be desired, the food is excellent, we went out on the topmost roof to sit. We were probably seventy-five feet above the waters of the Gulf and could hear the waves as they chuckled and sissed on the sand 150 yards away.

It was very restful and we were quietly enjoying our view without talking much when a large light object suddenly hurtled by us directly in front of our noses. We were sure it was a bird, a large one. Although we couldn't be certain, George thought it probably was an owl. So we watched for it.

Fifteen minutes later it repeated its flight, soaring by on completely soundless pinions as it followed the curves of the edge of the terrace where we sat. This time we had a better view as we were expecting it. George identified it with certainty as the creamy-beige plumage of a barn owl that was probably out hunting a juicy mouse or two for supper.

March 11, Veracruz, Veracruz

Yesterday we studied and collected duck food plants in the several scattered freshwater ponds between here and Alvarado. Two of the ponds are quite large and have an interesting birdlife upon them. We startled two little blue-winged teal exceedingly as we unexpectedly came upon them. They took off with muffled squawks and whistling wings.

There were small flocks of jacanas on two ponds, and with some effort, for they were nervous, constantly flying birds, we collected a pair. By doing so, we learned something we hadn't known before. At a distance, since their coloring is identical, there seems to be no difference to the casual observer between male and female. But in the hand a difference shows up clearly. The female is almost half again as large as the little male. And oddly enough she has the much larger pair of yellow horny spurs on her wings and a larger forehead shield. They have a beautiful green iridescence on each neck feather, in striking contrast to the rich auburn-chestnut of their backs.

We had a fine chance to observe the life of the little flocks. One group, which was made up of eight adult pairs and twelve immature birds, the latter probably the offspring from last summer's nesting, lived on the largest pond. Back and forth they fluttered constantly, looking very much like oversized butterflies as they flapped their wings slowly and landed with them held up together high above their heads.

At first we thought they might be polygamous with several females to a single male, as is the case with a number of birds. But by looking through the binoculars, we could clearly see the size difference between the sexes. And there was an equal number of males and females among the mature birds.

The two we collected were, their glands showed, almost ready for breeding. One pair of jacanas in the flock we watched kept entirely to themselves.

The male vigorously chased away any other males and all of the immature birds which came near them, raucously scolding in a loud tone as he did so. I imagine he was already defending his breeding territory.

We have a fine collection now of water plants of a considerable variety. And in one pond there were gigantic water snails, a full two inches across their dark brown shells, and with most unattractive slimy black bodies.

While George was wading in rubber boots at some distance from me, busily dredging up plants with a rake from the bottom of a pool for examination and I was at the car putting a pile of them in plant presses to dry, I happened to glance up at a slight noise on the fence beside me. There, not more than eight feet away, was a little ladder-backed woodpecker. He was industriously digging himself out a new home for the spring in a fence post.

Over and over again he braced himself with his stiff little tail feathers against the post. They acted just like a lever while he clung firmly to the edge of the hole with his long toes. Then in he strained beyond his shoulders and steadily chiseled away with his beak until I didn't see how he could stand the uncomfortable position for another minute. He was very busy and entirely preoccupied, and paid not the slightest attention to me.

Innumerable little yellow-rumped myrtle warblers flitted constantly over the surface of the ponds and landed lightly on the larger leaves of the water plants. One magnificent big bird, a Mexican black hawk [Common Black-Hawk] with wide white-end and mid-tail bands on otherwise all black feathers, and bright yellow legs and beak, observed us indignantly. His favorite perch apparently was the topmost limb of a dead tree in the very center of one pond we studied.

He eyed us fiercely with intense disfavor and screamed again and again as we got out of Chico. Finally, he decided that discretion was the better part of valor and flapped heavily off to a coconut palm not far away. There he stayed all the rest of the time we were working, observing us narrowly and occasionally giving a jarring cry, no doubt to hasten us on our way. He was beautiful with a proud, fierce carriage to his head, and a new entry for my life list of birds.

Today, since our waterfowl food collecting is finished, we decided to do a little general bird collecting for the museum. The grassy savannas have intrigued George every time we have passed. They look as though they should be perfect places for meadowlarks. But not until this morning did we feel that we had time to stop and investigate them.

We got up today at 5:45 A.M. As a reward I saw one of the most glorious sunrises that always makes me resolve to myself that I must see them more often, even if the hour is a horribly early one. But left to myself, since I dearly love to sleep late in the morning, I probably never would. The best insurance I can think of to make anyone regularly arise early enough to see sunrises is to marry an ornithologist who is carrying on studies or collecting in the field. Then it's inevitable, like death and taxes.

I looked out of our window over the Gulf this morning, and the sun was laying down a livid, molten golden path directly to me over the water. Overhead hung the sharp edge of but recently lifted dark gray fog. The sea was lighter steel-gray in color with ever-changing coral and amber surface streaks, much like the old-fashioned watered silk of my childhood hair ribbons.

The grassy savannas were very obliging. Having seen not a single meadowlark before during the several times we drove past them, this morning there was a pair drinking in plain sight at a little pool by the roadside when we arrived. Soon several males were singing farther back in the grass. We collected two males and two females, and they have turned out to be members of the race *mexicana*, which are definitely smaller birds than those of the highlands.

Then we went slowly on along the road to see what else there might be which the museum would like for study skins. A shiny blue-black *Dives dives* [Melodious Blackbird], that oddly named bird, was giving its bell-like song of three notes in a nearby tree. Each time he bobbed up and down like a tightly wound-up little metal toy. He has the ability, shared with a number of other birds, to sing two different notes at once. So all by itself it can sound like a well-executed duet. We collected it, two sparrows, a coal-black groove-billed ani with long curving striations down each side of its Roman nose of a beak, and a so far unknown (to us) little yellow warbler with small black eye patches.

We also saw various and assorted herons, innumerable flycatchers ranging from the large black-and-white stripe-headed derbys [Great Kiskadees] to the little nondescript and mousey *Empidonax*, Alta Mira orioles, a blue tanager [Blue-gray Tanager], several golden-fronted woodpeckers, a noisy flock of killdeers, Mexican grebes, Inca and Mexican [Common Ground-] doves, a large black and white hawk [Laughing Falcon] with a bandit's mask of black over both eyes, large green yellow-headed parrots, and eight little sharp-tailed green parakeets which flew directly over us, chattering shrilly to each other as they went. We had a glimpse of the crimson flash of an unknown woodpecker, like a spurt of blood as it darted past.

Group after group of great-tailed grackles were met in congress among the spreading branches of favorite big trees, noisily discussing the matters on their agenda. They constantly and restlessly hopped erratically from branch to branch like gossiping women at a county fair determined to miss nothing of interest.

George says it is customary for them to meet in large noisy groups like this each year before breeding and nesting begin. I don't know whether or not each "little woman" in a male's harem has a chance to discuss just what she would like in the way of housing facilities for the year. But that is exactly what it sounded as though they were doing, and in no uncertain terms either.

As we drove back along the highway, I noticed that in many of the coconut palm groves the unhealthy undulating trunks and branches of strangler fig trees were already creeping up and around many of the useful coconut-bearing trees. If they aren't soon cut away by someone, the more important palms will inevitably be killed.

George has made up our specimens into fine skins. We have been spending our last twilight sitting out on the roof terrace watching the full moon rise over the Gulf. It has been an evening of mauve, silver, and crowswing, a lovely placid sight.

March 12, Puebla, Puebla

We returned here from Veracruz today via Xalapa, or Jalapa, as it is sometimes spelled. This northern route has a more gradual climb back up to the plateau. And that is just what we needed with Chico. We first drove through many miles of beautiful plantations of pineapples, papayas, coconuts, bananas, and citrus, with a few scattered fields of manioc. Fine fat cattle browsed on extensive acreages of pasture, and it was a prosperous agricultural scene.

We stopped along the shoulder of the road in one place where George wanted to take a photograph of the countryside, and I stayed in the car. In looking idly around me, I was much amused at the actions of a small lizard on a tree trunk just by me. It was a coffee-colored little lizard, tastefully spotted with fine rows of a lighter cream up and down its sides.

What amazed me most was the amount of belligerency it apparently packed between its little ribs. First it moved down the trunk a foot or so nearer me. Then it lifted its body out from the tree and rapidly did push-ups with its two front legs threateningly, all of the while looking at me with an unblinking stare. When that had no apparent effect, down to the ground at the base of the tree it came. Then it ran over the bare ground almost to the car, where it repeated the maneuver almost without ceasing. And it was still doing so when we left. I suppose it is possible that it was mere curiosity, but it looked much too purposeful for that.

The lush tropical growth all too soon gave way to a very dry woodland. The latter reminded George very much of parts of Africa where he has traveled and collected. It has the same dry grayish cast to the vegetation and the scattered solitary flat-topped trees. The country soon became more rolling, and we crossed numerous limestone hills and then several steep gorges cut out by the river.

At Rinconada I had my first clear sight of the old Spanish colonial cobblestoned King's Highway, by which the Spaniards journeyed from Veracruz via Xalapa and Cholula to Mexico City. Rinconada may very well have been an overnight stop on the trip, as there is good river water and pasturage there. The ruin of an old Spanish building, a crenelated fortress tower with massive walls and narrow gun slits, still stands on the hill directly above it.

From that point on we often glimpsed the old Spanish road as it wound up among the hills. And we frequently crossed it on our newer highway. The Spaniards built well. In many places their ancient roadbed was raised on carefully laid rock foundations, or on small arched stone bridges well above the surface of the ground.

Along one long open shoulder of a hill, where trouble with the neighboring natives must have been prolonged and serious, they had built up eight-foot-high solid stone walls on either side of the road, fitted with intermittent gun slits. We could almost see the old springless carriages and the richly dressed attendant caballeros, guns in hand, and hear the clink of their horses' hoofs and the grind of the metal wheels against the old worn stones.

We passed nearby several tremendous old Spanish haciendas, now ruined, but with the delicate towers of their ancient chapels still standing. Several centuries ago they must have been the headquarters of well-producing, widely farmed lands. But today the fields are eroded and have reverted to brush.

Many farmers were planting their fields of corn today. They still use a very ancient and primitive method of sowing. A single man walks in a straight line through the field. Every few inches he sticks a tall sharpened pole into the soil and makes a hole. Then he drops a few kernels of corn in the hole and with one deft kick of his foot, covers them. He then steps ahead to the place for the next hole and repeats the process.

I should think it an extremely wearying occupation. And the rows and spaces between plants can't be very regular. But so they have planted their milpas since time immemorial, and so they doubtless will for generations to come.

More and more coffee plantations appeared as we climbed higher. Almost the entire countryside around Xalapa seems to be devoted to this one product. They grow the coffee bushes there under the shade of larger trees or banana plants, as is the custom in most of Central America too. But in Brazil and other parts of the world a greater yield is obtained by doing away with the shade.

The flowers along the road soon became as lovely as those of the Orizaba-Córdoba region to the south, and we saw many of the same species. We passed, growing wild along the irrigation ditches, clump after clump of creamy calla lilies, offering their waxed cornucopias to the blue sky above. No one apparently ever notices or picks them, probably because they are so common and grow so abundantly everywhere in the region.

We slowed down for a moment to look at a lovely garden in front of a humble one-room adobe house where the Señora must love her flowers very much. In a small handmade split-bamboo cage under the eaves of the house a yellow-headed parrot was swinging. When he saw us stop, being lonely perhaps, he started to sing to us. Over and over he trilled, in a manner guaranteed to give an inferiority complex to any living coloratura soprano. The flowers had recently been swept with a shower, and there was the soft, sweet, earthy smell of growing things in the air.

Xalapa is built at different levels along the side of a mountain, and it reminds me just a little of Taxco. The views from its balconies must be, on a clear day, superb. But heavy clouds had gathered by the time we reached there. And friends had previously told us that its climate is a wet, rainy one and that really cloudless days are rare.

Its name means in the Indio tongue "River of Sand," and refers to a large, usually dry sandy river nearby. It was a busy village when Cortés first marched through it on his way to conquer the Valley of Mexico. Many of the descendants of the original inhabitants still live in the region and carry on their lives with but little influence from the outside world.

If the day had been a clear one, we should have had from there another fine view of the Pico de Orizaba, and of a porphyritic mass named the Cofre de Perote, "Chest of Perote," which rises some 13,552 feet in altitude just to the west of the town. But they both remained hidden in thick billowing gray clouds, which rolled lower and lower as we continued to climb after leaving Xalapa.

Soon we came, at about seven thousand feet, into a beautiful pine forest. It was growing, of all places, on top of an old and extensive lava flow. The lava flow must be an old one since many of the trees are already seventy feet high or more. Our road wound through it for almost fifteen miles. It appeared especially interesting to me, as there were great clumps of luscious green mosses growing over everything . . . the lava, the scanty soil, and the numerous fallen tree trunks. So we decided to stop there for a picnic lunch. And I took the necessary few minutes to make a collection of all the kinds of mosses I could find. I got well over a dozen different species, which is a good many for one such odd ecological nook.

At Las Vigas the heavy fog was finally down to ground level. For twelve miles thereafter we crept slowly along, seeing nothing but the occasional edge of a sharp drop, or a tree looming eerily here and there as the mist momentarily thinned a little. Fortunately the traffic was light, and we went along in low gear with no mishaps. Our greatest fear, and probable danger, was that some truck driver, in that lighthearted way which is so common all over the country, might have stopped in front of us, full on the pavement with no light or other signal.

Just before we reached the town of Perote, the fog suddenly dissipated with the abruptness of a knife cut. We were at the same elevation and the country was exactly the same sort that we had just been going through, as far as I could tell. However, the sun became visible and the sky clear, and our spirits rose in direct proportion to the brightness.

After a little time we seemed to be going over a very large dry lake bed which varies from seventy-five hundred to seventy-seven hundred feet in altitude. At the town of Guadalupe Victoria there is still standing the tremendous old Spanish fort, Fortaleza de San Carlos, the "Fort of St. Charles." Its massive walls surround at least one and a half square miles, and it is striking to see, set as it is out by itself on the flat lake bed. Old Spanish watchtowers, pierced with the usual rifle slits, still rise upright at each of the four corners, and great buttresses support the walls. It was in fact a walled city, as within the outer walls we could see the roofs of a number of other buildings.

From there on, for almost thirty miles, the air was so filled with the thick ochre of the blowing dust of the lake bed that we could again see but very little. The soil is powder fine, and at this season the surfaces of the fields are mostly plowed bare earth. It would be far better if they were left planted with some drought-resistant cover crop, or at least covered with a trashy mulch of old chopped cornstalks to hold the light soil down. As it is, every vagrant breeze raises it to blow in clouds for miles and then finally to settle in high drifts against the houses and walls of some little town. That is just what it was engaged in doing today, with a particularly strong wind to push it along.

Shortly after we crossed the line between the states of Veracruz and Puebla, the road ran along the extreme edge of the crater of an extinct volcano. Only two yards or so to my right I could look down directly into the crater, which is now entirely filled with a lake of deep blue water. A few scaup were lazily swimming out in its center. At the nearer side a dozen women, on hands and knees at the water's edge, were scrubbing their laundry on large flat black stones.

Above the rim where we were, the wind still blew the irritating yellow dust in gusty, unpleasant drafts so that we breathed through gritty folded handkerchiefs. Below us, not even the ripple of a breeze stirred the surface of the cobalt water. The lake is about a mile and a quarter across, and round as a silver dollar with steep but uneroded sides. I wonder if the people who live nearby and use it appreciate what an unusual boon for Mexico they have in an almost inexhaustible source of water.

As we drove on along our way, we saw four more of the great walled hacienda-towns of the Spanish colonists. These are not as badly ruined as are most of the haciendas throughout the country where the Ejidatarios have taken over. Most of them have been senselessly razed stone from stone, as though by violent mob action, so that they are now entirely useless to everyone. But parts of those we saw today can still be used for shelter, storage, or even possibly as much-needed living quarters for a few farmers and their families.

The mighty, commanding peak of Orizaba appeared again to us briefly through a light haze. With the sun glancing off the great sheets of snow and ice that creep unevenly down its sides, it had a golden glow that reminded me of a George Inness painting. We finally joined the Puebla road by which we left that city for Córdoba, and we drove on along the valley, arriving in Puebla during the late afternoon.

Since we have long wanted to buy some of the old blue-and-white tiles for which Puebla is famous, and since the stores were still open when we arrived, today seemed a good time to get them. We left Chico safely parked at the hotel and took a bus to a factory which had been recommended to us.

It turned out to be located in a beautiful old Spanish colonial house, complete with the usual patio and tinkling fountain. The entire four walls of the patio for their full two stories are completely tiled in soft blues and whites and yellows. We soon found just what we wanted to outline the doors, windows, and balcony of the house we plan to build. And to George's great delight they had just one dozen of a smaller tile, each one of which has an appealingly original different bird painted upon it. We of course seized upon them and plan to mount them in a row across the fireplace underneath the mantel. These *azulejos*, as they call the Puebla tiles in Spanish, are all handmade. And they still use exactly the methods they started out with four hundred years ago when the Spaniards introduced the art into the city from Spain.

March 13, Puebla, Puebla

Today is Sunday. George is supposed to be, like all professional U.S. government employees, on a forty-hour work week. This should mean that Saturdays and Sundays are always free. However, in actual practice, except when we are in Mexico City, we almost never take off either of them to rest or to do just what we would like. There always seems to be so much that needs to be done and that we want to do in connection with our studies. So we keep going along with one day pretty much like another.

But we decided yesterday that we would take today entirely off. There has long been something that I have wanted to see in Puebla. And we have never before seemed while here to have enough free time at the hours it is open to do so.

According to the Constitutional Laws which were passed in Mexico in 1857, all the flagellating orders of nuns in the country were abolished and their convents closed. In 1934 when a census was being taken in Puebla, such an Augustinian order was, with something of a shock to local officials, discovered to be still operating clandestinely in the very heart of the city.

This Convent of Santa Mónica, as it was called, was founded in the seventeenth century. Its discovered building has now been made into a national museum and is open to the public. I have long been curious to see the physical arrangements of a place which could so successfully remain a secret for so long a time in a busy city like Puebla. So this morning we went to see it.

It is in the center of a city block of buildings, surrounded by modest homes and small shops, some of whose inhabitants certainly knew of its existence and helped to keep the secret. A simple home where two old people lived was immediately before it and the main street. They were apparently the ones who bought and brought in the necessary food to their own home and then passed it on to the nuns through one of several carefully camouflaged openings into the convent.

We were shown the formerly secret openings and how carefully and ingeniously they had been hidden. It was a large and completely organized establishment with several chapels, and separate quarters for the Mother Superior. Meditation, flagellation, and needlework were its three main occupations.

Its Chapel of the Viacrusis still has on display the crowns of sharp thorns and pointed iron belts with which the members tortured themselves. And in the basement there is a long, narrow, airless room with an "altar of death" which was kept at all times in total darkness, where the nuns meditated for hours at a time on their knees on the cold stone floor.

An especially clever arrangement was one whereby from an upper floor they could look through small peepholes, made on the outside to appear as part of the decorations, and observe the regular masses being said in the Church of Santa Mónica. The church was of course open to the public, and it bounded the hidden convent on one corner.

There were also two secret patio gardens. One was filled with fruit trees, and the noviriates took their exercise there. The other larger one was for the full members of the order and was beautifully arranged with flowers and formal walks. Their kitchen, completely covered even to the walls of the oven with blue tiles in the local *poblano* manner, is still just as it was found. Even the painted earthenware cooking utensils still hang by their nails from the walls.

Many of the rooms which have been converted into the museum are now hung with row after row of religious paintings, from this and other nearby convents. I must confess that in such quantities their rolling eyes and large quantity of gore are rather overwhelming. But the delicate embroideries of the time are well worth seeing. Especially so is a series of large tapestries in one room which were masterfully painted in excellent taste with oils on the finest of soft light brown velvet by the Cholula artist Rafael Morante.

We walked slowly back to the plaza, enjoying thoroughly the many sixteenth- and seventeenth-century buildings everywhere around us. A style of architecture seems to have developed here which I have never seen in any other part of the republic. Its Spanish and Moorish heritages are of course readily visible, but it seems to have an entirely different and original interpretation.

One of the most noticeable things about it is the number of unusual arrangements of the large flat vermilion building bricks. One common and very beautiful style in the city is that in which they have been laid in a vertical herringbone pattern with each block's outline clearly marked with a wide white border to make the simple design stand out. Often arranged at intervals among such bricks are single specimens or groups of the lovely old blue tiles.

These buildings have simple straight lines, and often carved niches still filled with some quaint little old religious figure, or an angel sculptured from richly carved stone. Many have balconies and window gratings, many of them masterpieces of hand-forged ironwork. Never does one know what the next block or street will bring into view.

After a late noonday dinner we dressed again in our field clothes and started out in Chico to visit an area of extensive alfalfa fields which extend southwest of Puebla, off and on, for about twenty miles. These fields are well watered, and George has long been sure that there should be meadowlarks in them. But he has so far collected none from there and needs to have a few in order to know what variety is present for the monograph he is writing. Also we thought that if by mischance we found none, we might do some general collecting for the museum.

We started out hopefully, stopping to listen and look at every likely field. But we were completely disappointed. We heard only two males singing, one within the city limits and the other directly at the doorstep of a little house where of course we couldn't collect them. Not another one did we hear or even see anywhere during the rest of the afternoon.

Finally, when even George gave them up, we paid more attention to the other birds around us. Eventually we took two small flycatchers and two sparrows, since they are almost always of interest to taxonomists and are often overlooked by collectors in favor of the more brilliantly colored specimens. Another specimen, a barn owl, was found crippled along the road and which we put out of its misery with some chloroform.

We drove on back to the hotel, and George has just finished making up our collected birds into museum skins. The barn owl has the softest feathers on its breast of any bird I have ever had a chance to stroke. Its dark brown markings on a creamy background plumage are really beautiful.

One of the barn owl's common names is "monkey-faced owl," George says, because of the great sunken, distinctly marked cheek areas of its face. These are doubtless why, in quick glimpses, it seems that an owl's eyes are so large.

March 14, Mexico City, D.F.

We left Puebla this morning for the last part of the trip back to the city. It was very hazy, and both Popo and Iztac were covered with gray clouds which drifted far down their slopes so that we could see only their bases. When we reached the town of San Martín Texmelucan, we were not more than a mile to one side of the boundary of the smallest *estado* in the republic, little Tlaxcala.

Since neither of us had ever been in that state, we turned off onto the road which leads in to its capital city of the same name. We followed it for about six miles, as George was interested in seeing whether there is anything of interest in the way of food and water to attract ducks or other game birds. We came across one reservoir which is large enough to have ducks on it in the height of the wintering season. So we shall inventory it another year before they have left, as they have already done this year.

The countryside is highly irrigated. Innumerable little silvery rivulets were shining like lacework against the dark earth as they trickled through the fields. Many areas looked like suitable spots for meadowlarks too. We hadn't the time today, but when we return again to check upon the duck population, I know we will also try to find out what meadowlarks there may be.

We are settled again for a few days in our favorite, comfortable hotel. There have been some very noticeable changes in the flowers and trees and birds in the city during the short time we have been away. To sum it up simply, spring is here. The jacaranda trees, which were dead-looking, leafless gray skeletons against the sky when we left, are now breaking out with their lovely sapphire-violet flowers and the smallest of light green lacy leaflets. Many of the streets in our neighborhood have been planted with them as ornamentals along the edges of the pavements. It is a very beautiful sight to look down a long row of them and see the handsome unusual color of their little tubular flowers against the deep blue of the highland sky.

The many rosebushes in the city have burst out, overnight it seems, into magnificent bloom. In the garden of the hotel the amaryllises are opening, each flower as large as a dinner plate. And the little borders of mesembryanthemums are aglow with a deep cerise light.

The song sparrows are busily nesting among the blue-flowered plumbago vines on the high walls. Some one or another of the males seems to be singing every minute of the day. They make a distinct competition for the rows of caged canaries and other birds that live in the open gallery along one side of the hotel.

For the first time I have had a perfect opportunity to hear over and over again the famed "whisper" singing which the song sparrows and a number of other birds occasionally do. While sitting in a garden chair looking at several darkly specked and striped little fellows that were hopping around not far from my feet this afternoon, I first noticed that one or another would stop abruptly periodically.

At first I noticed nothing more. Then it dawned upon me that although I heard nothing, at each stop the bird's throat was rapidly vibrating and his side swelling. By straining my ears the next time, I heard the most delicate of soft little trills, hardly more than the whisper of a far-off thing borne on a vagrant breeze. Over and over during the afternoon I heard it, while some little male stopped momentarily to express his happiness before starting out again on the business of busily searching for supper.

CHAPTER 8

Tehuantepec, Oaxaca, and Chiapas

March 18, Puebla, Puebla

We left Mexico City about noon today on the first leg of the longest field trip we will make in Mexico this year. It is to be down through Puebla, Oaxaca, the Isthmus of Tehuantepec, and Chiapas to as near the Guatemalan border as we can get. We cannot plan on hotels every night with the jeep's slow speed. So we have lightened the load considerably in Chico and have had a removable wooden shelf put in which extends the full length and breadth of the truck body. On it we have our sleeping bags spread out all ready to use if we are benighted along the way.

We mentioned our Playa de Cayaco experience to several close friends in Mexico City, both native-born and Americans who have lived in the country for twenty years or more. In reply we have heard a hair-raising set of authenticated tales of similar occurrences where the people concerned didn't come off as luckily as we, but were killed for their possessions. Everyone now tells us that they thought us very foolish when we said we would be camping out. But no one at the time went so far as to warn of the dangers.

Being used to camping through the Adirondacks, New England, and much of the Rockies in the United States, and George having done so many times in Canada, Africa, and northeastern Mexico, it never occurred to us that the rest of Mexico would be so different. But now we know. So we shall use hotels whenever possible and sleep in Chico only when it is strictly necessary. And never again in southern Mexico in the tent.

Today we once again crossed the flat lake bed of the Valley of Mexico and made the long beautiful climb up to and among the mountains. At the very top there is a little town, Río Frio, "Cold River," with half a dozen famous open-air restaurants, side by side along the edge of the *camino*. All the dozens of buses and trucks stop here on every trip, day or night. And many private cars do also, while everyone samples the delicious food the women prepare and serve.

Each restaurant always has in front of it a row of tremendous terra-cotta earthenware stew pots called *cazuelas*. Each one is of baked handmade pottery and is about two and a half feet across. They offer a variety of concoctions, bubbling interestingly over small charcoal fires. We always spend endless moments going from one to the other in the delicious torment of trying to decide which we shall choose for lunch that day. For by no stretch of the imagination can we ever sample them all.

At one side of the *cazuelas* on tables are big platters with piles of piping hot roast meats . . . pork, lamb, beef, chicken, and young goat, dishes of intricately made sauces, and heaps of hot tortillas. The latter are the thin rolled pancakes of finely ground white corn which are daily used in the place of bread by the majority of Mexican families.

Before long, after rolling down the far side of the mountains, we came to the small village of Huejotzingo. In the evening, we went into the center of the town from our hotel on the outskirts. We wanted to see the splendid cathedral and ancient tiled houses again. After sightseeing, we chose a quaint restaurant in which we ordered *mole poblano*. *Mole* is a sauce containing chocolate, cinnamon, and chile pepper. Mixed with meat and vegetables, it becomes a rural stew and is one of the national dishes of Mexico. Rich and delightful it is too with its original and piquant flavor, once one is used to the bite on the tongue.

March 19, Camp of the Silvery Oaks, near Huajuapan, Oaxaca

We left Puebla in good season this morning bound for the city of Oaxaca. But after some two hours on the way, miles from anywhere, and in the center of hot-desert country, Chico proceeded to have motor trouble. We stopped dead on a winding, twisting hill with not a foot of ground on either side on which to push him off the road.

George took various things off and apart in the broiling sun while I watched ahead for trucks coming down toward us, as likely as not on our side of the road. He finally diagnosed a part of the trouble, at least, as a defective new distributor. He couldn't fix it permanently but repaired it so that we could drive along a little while before the points began sticking again and the motor stopped. Then he again had to file and reset the points. In this fashion we proceeded all day without coming to a single town that had a garage with an auto mechanic.

We saw a number of birds along the way in spite of our troubles with Chico. Shrikes, lark sparrows, mockingbirds, mourning doves, and barn swallows were abundant, and several meadowlarks and white-winged doves were heard or seen. Just beyond Izúcar de Matamoros we had the good luck to see a group of band-tailed pigeons feeding, with flashing wings and tails, in some brushy growth.

Many of the hills and slopes we have seen today are very badly eroded. Numerous fields have a limestone outcrop that looks nothing so much as scrubbed chalk. It of course makes them perfectly useless either for pastures or growing crops.

Now we have come to the end of the flat hot country and are in the first scrub, oak-covered, mountainous terrain. We have driven Chico off the road and to the brink of a ridge so that we have this evening a series of endless views below and before us of alternating green valleys and mountain ridges. Winding down our ridge and twisting in and out for miles below is the glistening white ribbon of an old ox and burro trail. It was probably cut through the chalky soil by generations of Indian travelers hundreds of years before our road for wheels was ever thought of. We can see against its whiteness, two ridges away, the dark figures of poncho-clad muleteers like tiny ants inching along on some necessary journey.

Dinner is over. The sleeping bags in Chico are ready for us, and we are enjoying before we climb into bed the first hour of a rising full moon which turns the dusty oak leaves all around us to pure silver. I feel sorry for George. He has had a long hot tiring day with the balky motor. And he has the uncomfortable knowledge that tomorrow will be just as bad, and that we will reach Oaxaca only if he manages somehow to keep it going. How fortunate it is that we have had Chico arranged for sleeping inside.

March 20, City of Oaxaca, Oaxaca

In spite of our concern about the motor and the dirt and grease which George couldn't avoid getting over most of himself while working on it, today's drive has been a pleasant one. The country has not been so arid as that of yesterday, and scattered here and there along the route are fine small fields of wheat. Some of the patches are already a ripe gold while others still are green. Together they make interesting checkerboard patterns against the greater background of the uncultivated land.

The farther south we drove, the hillier the country became with deep valleys between the fine oak-pine forests on the crests of the mountains. Up and down and around we went for several hours until we finally wound our way down the last long curving mountainside and dropped into the valley of Oaxaca.

And glad we were to see it, as for some time this morning it didn't seem too likely in the future. However, each time Chico stopped, George managed somehow to get him underway again. Finally we rolled into the central plaza late this afternoon and already I like it so much. Just as we arrived, a splendid Sennett's white-tailed hawk appeared and continued its spiral above us for several miles as though he had constituted himself a welcoming committee of one for all gringos with ornithological interests, and was determined to do a thorough job of it.

Oaxaca has the comfortable, relaxed life and air of antiquity that we looked for and didn't find in Guadalajara. An old cathedral which was founded in 1553 rims one side of the main plaza. A long government building forms another side. The other two are made up of open-air arcades, or *portales*, which have small shops and tables and chairs outside along the sidewalk where one can sit and eat or drink and watch the rest of the city go by at its usual slow pace. With an altitude of 5,067 feet and a semi-tropical climate, the air is warm even in the evening, a delightful change from the nightly chills of Mexico City.

Before dinner we walked across the plaza from our hotel to a little outdoor café for a long cooling drink. We eventually stayed and watched the sights as they passed for an hour or more. Every moment the scene changed.

Townspeople sauntered along their way home from shopping or business. Little barefoot urchins made the rounds of the seated population bearing tiny cones of a watery sherbet which they sold for five centavos a portion. Indians approached to sell lovely handmade rugs and sarapes.

A blind woman led by a little girl came asking for a small coin. Shoe-shine boys one after another eyed our shoes beseechingly and asked hopefully, "Lustre?" And in the street, natives from all the neighboring villages and countryside trudged by on their homeward way from the marketplace in town, carrying their unsold goods in tremendous bundles on their heads and backs.

Over it all the soft balmy air settled. The heavy masses of the giant *Ficus* trees which are planted in the plaza slowly turned black against the sky. From someone's patio and far away the lovely fragrance of unseen flowers permeated the entire square. It was warm and peaceful and thoroughly enjoyable. We stayed on and on, talking or just sitting silently, until we had to run to make the dining room of our hotel for dinner before its doors closed.

Tomorrow Chico will be taken to a garage and, we hope, really fixed. Motor trouble in this country is no fun where, in so many areas, garage mechanics are few and far between.

March 22, Oaxaca, Oaxaca

I have investigated the market, several old churches, and countless little stores in the city during three days that Chico has been undergoing a searching professional examination of his innards. He should be ready for us tomorrow we are both happy to learn. George is anxious to get out on the small side dirt roads throughout the valley to see whether ducks, white-winged doves, or meadowlarks are to be found. Of course he hopes to find them all.

Oaxaca is becoming more and more charming upon further acquaintance. Its name is a Spanish corruption of the Indian name, Huaxyacac, meaning "Place Covered with Trees." It was founded in 1486 by soldiers of the Mexican emperor Ahuizotl, when they built a fort here on the banks of the Aroyac River while on their way from Mexico City to the Isthmus of Tehuantepec. In 1872 the name of the city was changed to Oaxaca de Juárez in honor of Benito Pablo Juárez, the Oaxacan who rose from poverty to become the president of Mexico. One of the churches I visited today, and I think my favorite so far in all Mexico, is the Church of Santo Domingo, which was founded in 1575. Its massive walls are almost fortress-like in style, and the interior is probably the finest example of Baroque decoration in the country. The walls and the great arched ceiling are completely covered with very heavily gilded ornamentation and sculpture which represent an enormous, many-branching tree. It twines its golden twigs and leaves in a network in all directions. Between the latter are the busts of many saints in polychrome, which become smaller as their height on the ceiling is greater until at the apex only their faces remain.

The ceiling above the main entrance is my favorite spot. I always unwillingly retreat from looking at it only when my neck vertebrae are shrieking with pain at the unaccustomed and uncomfortable position forced by looking straight up. Here there are thirty-four figures sculptured and set among the beautifully carved bunches of grapes of this most odd genealogical tree. It is a tremendously striking architectural spectacle and well worth the stiff neck that results from a really satisfactory study of it.

We have become very much attached to our little sidewalk café, and it is an ideal spot in which to spend the evenings. When we arrived at our usual table shortly before dark today, the grandstand in the park directly across the street from us was brightly lit and a uniformed band was softly tootling itself into key. Soon it started what we find is a semi-weekly concert. And excellent it is too.

We commented upon their skill and the classical selections they chose to play to an elderly Oaxaqueño who was slowly sipping his drink at the next table. Highly pleased at our praise, he explained that they still use a group of handwritten scores which were brought over from Germany in the middle of the nineteenth century by an earlier conductor. He thought that accounted for it.

Whatever the reason it is a pleasure to listen to them, and apparently the entire city thinks so too. Every bench in the plaza and rows of extra chairs were filled with the older folk. The younger ones stood or walked slowly around the square, enjoying a little mild courting at the same time.

March 23, Oaxaca, Oaxaca

We saw some very interesting backcountry today as we drove along the winding dirt roads . . . really only cart tracks in some places. One of the local industries is the growing of the castor oil plant, *Ricinus communis*. Its fruit is now just ready for the harvest. Small fields of the plant make intriguing color patterns with their big, deeply lobed green leaves and cherry-red spikes of prickly fruits against the intense blue sky.

Many girls and women were walking along the roads carrying back to their homes big piles of the cut fruit spikes, either on their heads or on the backs of small burros. We stopped at one little adobe house and asked permission of its owner to see what was being done to them.

In front of its house this family had built a spotlessly clean area in circular form, some twenty feet across. It was carefully paved with fitted rocks, and a three- or four-inch-high rock rim was built around its edge. Into this the fruits, still encased in their spiky husks, are spread out to dry. After a few days in the hot sun the outer shells are completely dry.

Then by running a short flat board over them with a rocking motion they split, and let the inner mottled bean-like seeds fall out. The useless husks are collected and removed, and the seeds are dried further. Eventually these are collected in sacks and are sold to the local owner of some small mill who presses out the oil and then either retails it to customers in the general vicinity or resells it to a larger company.

On many of the bigger farms campesinos were plowing with pairs of oxen which were teamed together by means of heavy wooden yokes bound to their foreheads. Many were pulling primitive wooden-pointed plows! We stopped to talk with and photograph one old man whose well-matched white oxen were wearing over their mouths very neat muzzles made of a heavy rope. I asked him why they wore them, and he pointed to the adjoining stand of castor oil plants. He said it was to keep them from taking bites of the plants as they passed, because they would be sick if they did.

Late this afternoon after work we drove the seven miles from Oaxaca to see the famous Mixtec and Zapotec ruins of Monte Albán. The natural setting of this extensive center of former civilizations could hardly be improved upon. A high mountainous ridge rises steeply and abruptly from the valley. On its very summit was built the group of temples and other buildings.

These cover a very large area and are still in excellent condition since most of them were made of stone. It was a well-fortified spot, and with its pyramids, subterranean chambers, and hieroglyphics, is well worth a visit. The greatest fame of these ruins probably is for the intricately hand-fashioned gold jewelry which was found in a number of the tombs when they were excavated. These tombs have been carefully preserved just as they were found, and an elderly Indio caretaker unlocked and flashlighted us through the one in which the greatest treasures were buried. The jewelry itself has almost all been kept as a single collection in the local museum in Oaxaca where it is on daily exhibition. It is of exquisite workmanship, and the designs are unique and most attractive.

March 24, Oaxaca, Oaxaca

We have been out twice today for birds, at daylight and again at sunset. It was a fairly successful day. We saw a dozen white-winged doves in a small grove of *guamúchil* trees (*Pithecelobium dulce*) and secured one. And George collected two meadowlarks in a nearby alfalfa field. He has just skinned them out, measuring, and examining them carefully to see which geographic race they belong to.

The drive itself was most interesting. We crossed the Atoyac River on the edge of town by a wide shallow ford and started off in a southwesterly direction. Our road was a narrow dirt track that apparently sees far more ox and burro travel than it does wheeled vehicles. But it was fun bouncing along and never knowing what unexpected sight the next corner would produce.

In one little town which rejoices in the odd name of Xoxo, it is difficult to know who was more interested in the spectacle the other presented, they in us, or we in them. We meandered on, blindly following the narrow road without knowing exactly where we were going until in a little valley we saw the *guamúchil* grove. With considerable effort, since they were extremely wary, we collected the one whitewing.

While in this grove we were just outside a town in which we could see silhouetted against the sky on the top of a completely barren bluff an enormous barrack-like building. It seemed to be extremely old from its architecture. So we went on and found the town to be Culiapán, and the building a famous Spanish convent, of course abandoned and in ruins now.

The town is even more famous historically for something else. It is said to be the original home of the cochineal industry which was at one time a very important and flourishing business. The Aztecs were probably far ahead of the Europeans in the art of dyeing cloth at the time of the Spanish Conquest. By the use of the various plant parts, soil, and such chemicals as niter and alum, they knew how to obtain a very large number of beautiful colors.

But one of the most difficult colors all through history has been the obtaining of permanent colorfast red-purples. This is very probably why in the days of the Greeks and Romans purple was a royal color and reserved for the highest nobility. The Aztecs had beautiful and permanent red-purple dyes which were first discovered at some unknown date by the early Mixtecs or Zapotecs who lived in what is now the town of Culiapán.

They found that the female cochineal insect (*Coccus cacti*), or *nochiztli* in their language, when placed and fed on the leaves of the nopal cactus and later collected and killed and then dried, would give a rich lasting dye of a particularly lovely red-purple color. This was used for the robes of the kings and Aztec nobles in the country. Of course after the arrival of the Spaniards, the secret was taken to Guatemala and to other parts of the Spanish empire.

It is interesting that even today in a single small mountain village east of the city of Oaxaca, which can be reached only by difficult trails, the natives raise silkworms. From the silk they make singularly lovely raw silk rebosos which are always dyed with the beautiful cochineal color.

They make but a few annually and only bring them down to sell in Oaxaca when the spirit moves them. But if a person is fortunate enough to see one, as we did in Ernesto Cervantes' fine shop there, it gives one some slight idea of the skillful and extraordinary workmanship of which the unlettered Indians of former days were capable. Even today we cannot satisfactorily reproduce this color with all of our knowledge of machines and chemistry. No company has yet been able to produce the exact and valuable cochineal tint in a colorfast dye which will not change with either sun or time.

We retraced a part of our path and then took another narrow road south through a fertile valley in which agricultural crops were growing. Finally we rolled into the village of Zaachila. It showed signs, we realized, of having been at one time a prosperous if not important town. However, many large buildings there are now unoccupied and slowly falling down. But it was not until we were back in our hotel in Oaxaca and casually chatting about where we had been during the day that we really learned about Zaachila.

It was originally the capital city of all the Zapotec empire and therefore, of course, of great importance. It is there that the beautiful and original Danza de la Pluma, or "Feather Dance," is still performed on fiesta days. We have seen one of the magnificent costumes in the museum in Mexico City, and they are striking indeed. The tremendous fan-like headdresses are made of gaily colored feather plumes, as are the tufts and bands that are worn on the arms, legs, and bodies of the dancers. Now someday I want to be in Zaachila on a day when the dance is being given and see it for myself.

I had an amusing glimpse of home life in one tiny village we drove through today. In the doorway of a small thatched hut with a hard-trodden bare earth yard in front of it, two women were sitting on the ground. The older was industriously combing and searching for possible small livestock in the long hair of the younger. The latter held upside down on her lap a turkey hen, the side of which she was slowly and regularly stroking as though it were a lap dog. The turkey calmly and serenely remained upside down, apparently thoroughly enjoying it all.

Tomorrow we start off for what is to us, *tierra desconocida*, "unknown land." We still have not been able to find out much about it that promises to be accurate. As we have discovered before, we really won't know just what conditions are until we are on the scene ourselves. Chico ran well today during the two trial trips, so we fervently hope that we have had the last of our troubles with him.

March 25, Tehuantepec, Oaxaca

We have come so far with no trouble, and on a splendid paved road all of the way. After leaving the Valley of Oaxaca the country becomes more and more desert-like and hot. Yet in the middle of the worst heat of the day, in a completely arid section where hunger drove us to stop for lunch, a white-winged dove not far away was in full blasting sunlight lunching on the juicy fruits of an enormous candelabrum cactus. George was so interested in the sight that it was with difficulty that I got him to put his binocs down long enough to finish lunch.

Having always read of Tehuantepec's beautiful women, and knowing that the city is situated on a low isthmus of land, I expected a romantic jungley setting with dripping vines and festoons of exotic orchids, and more or less permanent moonlight in the background. My disillusionment has been intense to find it a drab mud-colored town, in this season of the year completely flowerless. Everything in sight is stark brown and hot to the touch from the terrific heat.

The women are tall and stately, and their long skirts and low-cut blouses are very flattering. But their usual trials of year-apart children detract considerably, in my mind at least, from the romantic ideal. It is only fair to say that in the rainy season the town must be more presentable since some flowers certainly are then in bloom. However, in the dry season any Chamber of Commerce would be hard put to find copy for a single advertisement.

We had been recommended to a hotel, but we had great difficulty in finding it. No one could tell us where it was, even when as it turned out we were but a block away from it. I can now readily understand why.

Our room for the night is tastefully furnished with two wooden frames having a canvas slung between. On the canvas is a paper-thin dirty cotton pad and one sheet which reaches three-quarters of the length of the canvas. This is a bed. We also have one chair, two wooden pegs in the wall on which to hang our clothes, and a small rickety table which bears a pail of water.

The one bathroom for the hotel's entire clientele, and there are no facilities for bathing or showering in it, is a tiny closet-like affair with no window or lighting, and a door which closes only uncertainly. It is built for some unfathomable reason on an inner patio next door to the kitchen. Since it has no running water and temperatures are high, the result has to be experienced to be fully imagined.

Supper tonight was served in the main entrance room of the establishment. We walked carefully to avoid holes in the floor, and a single naked light bulb on a wire let us see all too dimly what we were eating. There are mosquitoes in quantity and no screens. We anticipate a lively night since our DDT bomb can't cope with the entire hotel.[1] There is no way to shut off our room from the rest of the building to spray it alone as the door reaches only from one's shoulders to knees.

March 26, Tuxtla-Gutiérrez, Chiapas

The night was lively, but not quite in the manner I had anticipated. Something in the amalgamation served me as meat pie at supper last night was apparently considerably too ripe. I suffered as a result. George took one look and said he didn't think he was very hungry. But I was, so I ate it anyway, and I have regretted it ever since. All today when I particularly wanted to see the vegetation of southern Oaxaca and northern Chiapas I have been *hors de combat*, not caring at all where we were going or what we were seeing so long as it led rapidly to a bed.

Chico picked high noon at the beginning of a particularly long, winding, narrow mountain road to stop dead again. George worked and worked on the motor, but at best it would go only a few miles and then stop. The few villages we coughed in and out of had neither a garage nor a mechanic. Everyone we asked assured us that there was none of either before reaching Tuxtla-Gutiérrez, our destination.

After several hours of this, when the jeep finally seemed stopped for good, a Mexican workman in an ancient rattletrap Ford pickup truck came along and parked by us. He knew a little about jeeps, but not much. However, he did his very best to be helpful. We were more than forty miles from Tuxtla, and he was on his way there. Finally very ignominiously we attached Chico by a rope to the rear of his Ford and were hauled up hill and down dale, when George happened to try the motor at one point for no good reason, and it started again. So we made the last few miles under our own power, our good Samaritan having gone on after seeing us safely start.

I anticipated another Tehuantepec. I was almost overwhelmed to find Tuxtla a charming and beautiful little city with a brand new deluxe hotel on its outskirts. We have one of a series of small cottages which are furnished in comfortable last-minute American style and placed in the middle of a flower-like garden.

A swimming pool and bath house are about two hundred feet away as is an outdoor bar. There is a ping pong–billiard room, and the handsome glass-walled dining room is serviced by white-jacketed waiters. And the beds have innerspring mattresses and our bathroom has a shower! Verily Tehuantepec seems a thousand miles away, and as far as I am concerned, it can stay there!

March 31, Tuxtla-Gutiérrez, Chiapas

If possible, Tuxtla even improves upon acquaintance. It is still in that lovely rare state of being completely non-tourist conscious. This is for us a wonderful change after so many towns and cities where the standard, when a U.S. car parks, is for all the small boys in the area to launch themselves on and over it, and for all the beggars to gather where they must be pushed through before the car's occupants can leave.

In Tuxtla at this date a tourist is about on a par with the occasional Indio from some one of the mountain tribes nearby who wanders down, bobbed hair, dressed in a single garment, and with a spear for armament. We are I imagine, tourists, Tzotzils, Zoques, Zapotecas, and others, all lumped together as undoubtedly different from the Tuxtlans, but not people therefore to be stared at rudely or bothered. It is a very fine and civilized approach.

The central plaza is filled with flower gardens and trees, and benches are provided where one may sit and watch the affairs of the town pass in review. Several outdoor cafés serve cold *refrescos* (sodas) of all kinds to those who prefer to sit and sip to just sit.

I have been basking in a few days of almost utter laziness while George first spent two days getting the local jeep agency to go over Chico again, and then went down by bus to Arriaga and Tonalá on the Pacific coast. These towns are, at this season of the year, hotter than the proverbial hinges of Hades. Whether he would find even a hotel in existence at Mojarras, his ultimate destination beyond Tonalá, was a moot question. So it seemed wiser for me to stay here.

I packed bread, peanut butter, cheese, and fruit in a knapsack for him, so at least he can eat whether there is a hotel or not and even though he has to roll up in a corner of a native hut to sleep. He has been anxious for some time to study from the ground several coastal lagoons in that area which had very large waterfowl concentrations when he surveyed them from the air a year ago. And too, the white-winged dove migration is said to come over there each spring, so he wants if possible to spot it accurately and estimate their numbers.

Shortly after our arrival at the hotel we made the acquaintance of a yellow-headed parrot, Lorita by name. *Lorita* in Spanish is "little female parrot." She has an exceedingly strong personality and has become one of the attractions of the place for me. Lorita has a consuming passion for all ladies and just as confirmed a hatred for all gentlemen. She doesn't hesitate a second in showing it.

We discovered this almost at once after breakfast the first morning. We were working in our room when we heard her trudging down the covered corridor-like porch. She was clicking her long toe nails on the tiles and audibly announcing in excellent Spanish, "Who is talking? Lorita is talking. Lorita is coming." Since we paid no attention to this, she then knocked several times on the bottom of our door with her beak and continued to do so and talk until we opened it.

In she walked, chuckling and gurgling to herself with pleasure. She first made a brief but well-aimed foray against George's shoes to show that she would brook no nonsense from him. Then she ambled across to my chair which she climbed until she could sit upon the top of its back. She stayed for an hour or more, going through her repertoire, singing, whistling, displaying her beautifully marked red and blue wings, and being generally entertaining.

She has continued her visits about three times a day, always going through the first ritual of walking down the corridor and knocking on the door. She has been unfailingly gentle with me and has loved to have me sing with her. I have become quite fond of having her around. With George, however, it is a slightly different matter.

Try though he may to win her over, she persists in spreading out her wings and striking at him for absolutely no reason whenever the notion enters her head, which is often. She always keeps her beak closed and has never bitten him or his clothing, but it is a distinct blow to his pride as an ornithologist to encounter a bird who so clearly will have none of him. Each time he retreats after making an effort to be friendly, Lorita screams like a hussy with fiendish glee and triumph at the top of her anything but subdued voice, which of course doesn't help his feelings any.

Most of the time during the days he has been here, he has been busy working on bird skins or notes at a small table in the center of our room. Of course he hasn't had too much time to pay to Lorita. So it has pleased her greatly to quietly climb down at intervals from my chair where she usually sits when visiting us, pigeon toe over to where he is unsuspectingly working, and with an ear-splitting shriek and widespread wings, launch herself upon his ambushed feet.

This done to her satisfaction, but without a bite having been inflicted, back she always self-importantly marches to my chair, talking to herself with vigor. Then she climbs up it again to be her sweetest and most affectionate self with me.

As a favored friend I of course have found her strong-willed rejection of George highly amusing. But I must admit that if I had been treated as he has been by her, I probably wouldn't have found her nearly as entertaining as I have. But with his usual good nature he always welcomes her when she appears again, invariably hoping that perhaps this time she will be friendly to him too. If I were he, I am afraid that I'd feel more like wringing her handsome neck.

April 3, Tuxtla Gutiérrez, Chiapas

In the market and around the central plaza in Tuxtla, the Indian women carry many of their purchases on their heads in beautifully painted old gourd dishes. Some are the largest that I have ever seen, being almost one and a half feet in diameter and four or five inches high. The patterns are usually of large stylized flowers of various colors against backgrounds of a single contrasting color.

Some of the oldest gourds which have been used for many years are really lovely. The colors are very soft, and the workmanship is much finer than that of the more recently made ones. The next village to Tuxtla, Chiapa de Corzo, is where most of them are made today by a number of the local families. They are then taken around the countryside and sold at nearby markets.

George is back from his trip to the coast after fair to middling success. In one small village we drove through recently on a short field trip, we passed orchard after orchard of *jocote* trees (*Spondias* spp.), the first I have seen since leaving Oaxaca. The fruit is a yellow drupe of elongated small plum size and shape. Since it always appears on the bare branches of the spreading low trees before any leaves are present, it gives them a topsy-turvy look. They are delicious in flavor and must provide a welcome variety to the usual local diet. I am surprised that they aren't grown here more widely in the lower hotter altitudes in Mexico, for they are very common in Guatemala.

On the outskirts of the same village we saw a family making a slender rope. Apparently its manufacture is, in this section, a small family industry. Behind their one-story adobe house, and extending for perhaps a hundred feet down a slight slope and fifty feet in width, was their patch of fiber-bearing agave plants. Since none of them was in flower, I couldn't tell whether they were sisal or henequén or some one of the several other agaves which give a usable fiber.

In the process of rope making, the outer leaves of the plants at the proper time are cut off and put to soak, usually in stagnant water. This makes the softer tissues rot away and leaves only the hard vascular bundles which make up this type of fiber. At exactly the right time and just before the fibers themselves start to be affected by the rotting process, they are taken out and hung over lines or fence posts to dry in the hot sun.

We arrived on the scene just after this stage had been completed. As we watched, the mother and daughter of the household were fluffing up the dried fibers and arranging them in rough hanks of what seemed to them a suitable size. Father and his son, the latter aged possibly ten years, were out along the roadside making the actual rope in a primitive but skillful manner.

Somewhere they had obtained a small metal wheel that turned by a handle at one side. They had fastened the wheel to a wooden stand so that it came about waist high. Fastening the ends of some of the fibers in a hank to the wheel, the son was engaged in turning it at a constant speed while Father walked backward along the edge of the road. As he walked, he slowly pulled out and fed the fibers from the hank until they were all twisted into a single thin rope.

After asking their permission, which was politely given, I took a series of colored photographs of the various steps they were then carrying out. Afterward a few cigarettes given to Papá apparently left everyone completely satisfied. We were sent on our way with a chorus of, "Que les vaya bien," literally translated as, "That it may go well with you."

April 7, Tuxtla Gutiérrez, Chiapas

We have been spending several days studying the migration of birds, particularly white-winged doves, quite a number of which we have discovered to be in a valley several miles north of town. They are a wary lot and we have had quite a time stalking them in order to collect several. We want to see whether they are of the eastern or western races and whether they are local or migratory birds, things which no one yet knows about those found in this vicinity.

We went out there after whitewings today. In order that the shots of one of us would not ruin the stalking of the other, George took one side of the valley as his territory and I the opposite. My usual hunting costume in such a hot climate consists of an old faded blue cotton dress, since it is much cooler than slacks, and blue canvas sneakers minus socks. Around my waist on a wide leather belt are always slung two bandoliers, one filled with .32-caliber and the other with .410 shells. A long hunting knife dangles at one hip.

Binoculars are around my neck on a strap so that I can definitely identify a bird before collecting it. Over my back hangs a khaki knapsack in which to put them afterward. And banging against my stomach on a string is a short ramrod, as empty shells often stick in the auxiliary barrel and something must be handy to poke them out with before I can reload. Over one arm I carry my .410 shotgun.

After a few dirty streaks have gathered on my face and arms, and I have completely disheveled my hair by pushing through bushes and small trees, I imagine I appear about as tough a character as any woman, gun in hand, can look. While in this general state this morning and half a mile or so away from George, I was stealthily advancing up a narrow trail in dense brushy cover intent on an elusive bird song ahead of me. Suddenly I came around a curve and found myself face to face with an elderly Indio.

He was dressed in the usual white pajama-type suit of the region and carried a well-sharpened extra-long machete. I think it was really a toss-up as to which of us looked more villainous, or was more startled at the other and jumped the higher. Then for perhaps a full minute we eyed each other warily for the next move.

Finally simultaneously we said, "Buenos dias," sidled past each other uneasily on the narrow track, and walked as rapidly as we could go on our separate ways. I couldn't help wondering later just what sort of a tale he made out of his experience by the time he reached his village, because native women do not go off hunting with guns in Mexico. And he may never before have had a close look at an Anglo woman. I must have appeared to him like something literally out of this world.

I had planned to make a fine plant collection in general, and one especially of grasses, from Chiapas since the National Herbarium in Washington, D.C., is interested in specimens from here. To my intense disgust, however, I find that in the dry season now there are normally but a few plants in flower in this area. And specimens must be in either flower or fruit to be identified with certainty and of any use in an herbarium. On top of that, this year's rainy season was of very short duration. So the countryside of Chiapas is bare brown earth and brownish-leaved shrubs, almost completely flowerless except for a few common roadside weeds.

Even the birds are fewer in number than we had expected them to be. We have gone for hours through woods along the one large river in the vicinity where the vegetation is dense and where one would expect lots of birds to be, and have not seen half a dozen.

Yesterday we drove fifty miles beyond Tuxtla Gutiérrez to San Cristóbal de las Casas. The road is a little on the rocky side but perfectly suitable even for a passenger car. It is fifty miles of continuous mountain driving with ever-changing beautiful views.

Just as we were starting the steep grade up the mountains from the valley in which Tuxtla and its neighbor, Chiapa del Corzo, lie, I happened to see sitting side by side on a low shrub by the edge of the road two lovely birds of metallic blue color. Each had a most peculiar tail which included two long bare feather shafts ending in small lollypop-shaped tufts of the original feather material. They were mot-mots and we collected one for the museum.

The males always have the very odd habit of carefully and neatly stripping with their bills all of their two long middle tail feathers except for the little racquet-shaped remnants at the very ends. Could it be vanity carried to an extreme, and do they adorn themselves in this way for their female's benefit? Perhaps though, it isn't any more of an extravagance than milady's annually new Easter bonnet. And it may well serve the same purpose, in reverse.

The Indios along the way are still so unused to strangers that many of the women turn their backs and hide their faces when a car drives past them. Most of the men we saw were wearing handwoven white cotton track shorts and a narrow-brimmed straw sailor hat with colored ribbons hanging down the back such as little boys wore at the turn of the century in the eastern part of the United States. Their unusual sandals have high leather backs which I have seen elsewhere only as illustrations in the paintings or stone carvings of the ancient Mayas. They are undoubtedly a rather primitive group still living in many ways much as their ancestors did centuries ago.

Tuxtla is about 1,775 feet in altitude, and Las Casas lies over 7,000 feet. During the climb we passed several areas of fine pine forests and saw and heard many more birds at the higher elevations than we had previously in all of the lower part of Chiapas.

It was an odd experience to see the collared robin [Rufous-collared Robin], a bird that acts much like our familiar robin redbreast at home, but is instead entirely black in color except for a golden-buff inch-wide collar around its neck. It does, in fact, belong to the robin subfamily. A number of brilliantly plumaged blue jays, looking like enormous metallic butterflies as the light was reflected from their feathers, flitted in the pine groves as we drove through. We stopped several times to listen for the songs of white-winged doves, but heard none until we reached lower country.

Las Casas is over four hundred years old, and with its ancient plazas, churches, and other buildings, it looks like the sort of town we would love to stay in and explore. Unfortunately we have no spare time to do so on this trip.

We have spent two days now preparing as museum skins the bird collections we have made in Chiapas, writing up notes, and doing other necessary chores. Tomorrow we start again on our way to Mexico City. It has been a most pleasant stay in a bewitching town among friendly and interesting people. I hope so much that the influx of tourists when the road is finished to Guatemala City, as it surely will be soon, will spoil neither of them.

April 8, Camp of the Chachalacas, near Níltepec, Oaxaca

We left Tuxtla at a very early hour this morning and by late afternoon had arrived on the main road north of Zanatepec. We spent the last several hours watching constantly for migrating white-winged doves anywhere on the horizon. No one has known just where they travel after leaving the Arriaga-Tonalá section until they reach southern Veracruz.

The question has been, do they fly only up the coast, or is it their custom to cross inland and along the sides of the more or less continuous line of low mountains, or do they change back and forth? From what is known of the habits of other migrating birds, George thought that they might come inland and follow the base of the mountains up to and across the Isthmus of Tehuantepec. Hence any place where the main highway comes close and parallel to the line of mountains in this area theoretically should be an excellent spot for us to stop and look for them.

This maneuver was somewhat complicated, however, by the fact that whitewings are sometimes rather choosy as to just when they migrate. Their favorite hours to do so are in the cool of the early morning when neither people nor the sun is out in any force. Then they are very likely to stop for a drink at some stream or pool which may be in the vicinity, and afterward to roost in nearby trees during the heat of the day. They remain there quietly until afternoon when they make short flights to feed on whatever fruits or seeds may be available, after which they may start out again on their typical long, low lines of migrating flight. Toward dusk they roost again for the night.

So in order for anyone to see migrating whitewings, he must be in the local spot the birds choose to go over, usually in the early morning or late afternoon. Otherwise none may be seen.

It seemed to me a little less remunerative than hunting for a needle in the well-known haystack to try and find the unknown place in all this side of the isthmus where the road and their flight line might coincide. However, by a careful study of the topographical maps of the area, we picked one such place and reached it at the proper time this afternoon.

We set up our work tables off the side of the road under a gorgeous spreading tree which was loaded down with brilliant crimson globular fruits, each looking like a small juicy plum. They lay thickly on the ground around the tree giving off a delightful fragrance. But their taste must be very bitter or their fruit poisonous, as neither bird nor Indio in the region seems to come near them. Any fresh fruit in this arid territory is a treat that would certainly be raced for by anyone if it were edible.

I sat for over an hour with binoculars glued to my eyes constantly searching the different sectors of the sky, while George skinned and made up as museum specimens the birds we had collected during the day. But no whitewings appeared.

That is one major and lovely difference between having as a life work botany or ornithology. I occasionally rather unkindly remind George of this when he is tired after a long day in the field and is still faced with many more hours of work in order to prepare the collected birds before they spoil in the heat. Plants can be tidily tucked between newspapers to dry, and there they will stay safely without much attention. That is, unless the climate is very humid, of course. Also they stand still when one wishes to collect them and don't have to be stalked endlessly, usually through the very densest, spiniest thickets in the neighborhood.

George is, fortunately for me, blessed with a wonderful disposition. So the retribution which I no doubt richly deserve for such teasing at such a time does not descend upon me. He merely steadily goes on skinning birds, doing no more than perhaps showing a canine tooth lightly in mock anger.

Finally when my arms and legs were thoroughly tired and sunset was near, and I was quite disillusioned about the whole thing, groups of from twenty to seventy-five whitewings began to appear overhead. They flew in from the south and then along the mountains not far away!

Most of them dropped down for a brief drink in the river that crossed the road there too. Then they rose again, in any direction apparently, but all steadily turned in a wide circle until they were once again heading northward.

We stayed until the last ones had gone through for the night, George estimating and I writing down the numbers of birds in each flock. Then it was high time for us to think about a place for us to spend the night. We make it a rule now never to camp overnight within sight of the road, as that is simply asking for attention from passersby. But sometimes we have to drive a good many miles before we find a place where Chico can be driven off the main road and into enough vegetation to camouflage him and give us the privacy we want.

Tonight we hit just such a difficult area. We had to drive until only the last rays of the afterglow were left before we happened on a splendid place in a grove of large trees. I have never heard such an outpouring of bird songs as was still going on when we got out of the jeep. Chachalacas [White-bellied Chachalacas] easily led the chorus. It sounded as though there were hundreds of them in a circle entirely around us and only a few yards away. Actually, George says, there were probably not more than six or eight chachalacas. One of their outstanding characteristics is the ability for a few to sound like a town meeting in full cry. But the camp has been named with neatness and dispatch.

We hurried to finish a quick cold meal before pitch darkness could overtake us. But it arrived before the end of washing the supper dishes. We are more than ready for our sleeping bags on Chico's shelf tonight. And it will be noisy visitors indeed who manage to wake us.

April 9, Salina Cruz, Oaxaca

We continued today toward Tehuantepec, and having my eyes open and being in good health and spirits, this time I saw the countryside. As we dropped down into the lowlands of the Tehuantepec isthmus, the wind picked up steadily until it was hitting us sidewise with hard, gusty blasts. We were glad that Chiquita had been left in Mexico City and was not perched on top to present even more of a surface to it.

The local winds in this section are called *Tehuantepequeras* and are infamous all up and down Mexico and Central America for the suddenness with which they blow up and the strength to which they rise. For the wind to blow 100 to 125 miles an hour during one is not particularly unusual. We fortunately hit nothing like that, but it was strong enough to satisfy our curiosity completely.

Shortly before reaching the city of Tehuantepec, which means according to some authorities "Hill of the Tehuanas," and to others "Mountain of the Man-Eaters," we came to an extensive natural grass savanna. It is an isolated area which is separated by wide mountains from savannas to the north and south of it. As we were driving along through it, we suddenly heard the familiar sound of a meadowlark.

George of course parked Chico immediately, spotted the singer with his binocs, and eventually went after him with the .32. He collected another one also to see whether those found here are of the variety first discovered many years ago in a more eastern part of Chiapas.

We were both happy to be able to bowl right through Tehuantepec this time. If I may correctly call "bowling" creeping in low gear down a riverbank, crossing the wide, uninteresting river with a ford because the bridge was washed out years ago and never replaced, and crawling up the far bank also in low gear. We turned south at the edge of town and continued for eleven miles over a gravel road to Salina Cruz.

Salina Cruz is Spanish for "Saline Cross" or "Saline Sorrows" and may well originally have meant either one depending upon the conquistadores' experiences there. In past days it was a major seaport on the Pacific coast of Mexico. Now it has a sad, nostalgic air as though it were dreaming hopelessly of the wonderful days that were. Its blank-windowed big buildings stand empty for blocks along the main streets, and the enormous warehouses on the docks rust quietly away. The one thing that keeps it even half alive is that it is the southern terminus of the only railroad across the isthmus from the Atlantic to the Pacific Ocean, which was opened to traffic in 1907.

It has been periodically considered as a likely place for an inter-oceanic route since it is the narrowest part of the continent north of Panama. Whether anything will be done to open another shorter way across the continent for commercial ships in time of peace and warships in case of war remains to be seen. At the present time the railroad is functioning with a single daily train in each direction between Salina Cruz and Coatzacoalcos, or Puerto Mexico as the latter is also called, 192 miles away on the Atlantic coast.

We have found an old-fashioned but perfectly comfortable hotel with private bath and running water. We had a simple but edible dinner there this evening, so we feel very smug and pleased with ourselves at having so bypassed the rigors of the Tehuantepec hotel.

April 10, Salina Cruz, Oaxaca

We have spent most of today driving along all of the backcountry dirt roads we could find in the vicinity. We wanted to scout out the lagoons for ducks and to see if any of the migrating white-winged doves come up along the coast here. There were a few ducks, but we found not a single dove. We did see and collect a number of other migrants, including an incredibly colored Leclancher's nonpareil [Orange-breasted Bunting]. This bunting-like little bird boasts a bright yellow face, gaudy kelly-green head, blue back and tail, yellow underparts, and flaming orange breast. He looks as though he had been dipped in and out of a succession of paint pots by a mischievous deity.

There were also a white-fronted dove, a red-billed pigeon, a lovely but still unidentified trogon, and several flycatchers. One little *Empidonax* led me a merry chase when I tried to collect him, and I never did get within range. Flocks of small and long-tailed parrots flew over us off and on all day, usually chattering stridently as they went. They seem to be among the most sociable of birds and always are talking to one another, though they obviously can't all be females!

While George was in the brush retrieving a trogon he had collected, I idly walked along a nearby sandy track through the trees. Suddenly there ran out in front of me a [Lesser] roadrunner. At first he didn't see me. When he did, he flipped his tail and took off and out with the velocity of a miniature sailplane.

We saw many lizards of all sizes from tiny two- or three-inchers to the three-foot edible iguanas. One of the commonest is a six-inch one light gray in color, which has the ability to pop out a bright coral-peach throat balloon at an instant's notice. It does so constantly when in any way startled. Several times today I found one near me on some tree trunk, leaning far out from it and rapidly puffing in and out at me. Since of course I showed not the slightest fear, he usually tentatively popped it a few times more while eyeing me coldly, and then ran skittering away.

One winding, very sandy road led us to a tiny thatched village on the coast where the villagers stared in amazed surprise at the sight of Chico. They seemed to live principally on fish, as several nets were hung up to dry and others were being mended by a few elderly men. Aside from coconut palms and a number of chickens, we saw no other food of any sort being raised.

A vista in Oaxaca showing productive croplands in the foreground. In the distance are steep, higher-elevation brush and forests that farmers have cleared for marginal cropland, using the *tumba y quema*, or "slash-and-burn," practice. Such farming, usually for corn production, is short-lived and often results in severe loss of topsoil and erosion, evident in the picture. (George B. Saunders files)

The children were mostly bare-bottomed and adorned only in completely inadequate little white shirts that reached about to their lowest limbs. They were obviously not used to strangers, as they just gazed at us with open mouths while we backed Chico around in the tiny clearing at the end of the road. They never so much as pointed or cried, "Mira," "Look," to each other as is the usual reaction among the small fry of the little villages we visit.

Their mothers, a few of whom gathered silently to watch us from the doorways of their huts, didn't seem to have much more initiative then the youngsters. I suppose their lives are dull and monotonous in the extreme. But they rather looked to me as though they were too.

I had plenty of time to examine a typical local woman as George was slowly backing Chico back and forth in the tiny space at our disposal. She had a flat nose, high cheekbones, and wide nostrils. Her hair was long and black, and braided into two thick ropes that dwindled into rag-tied ends. A too-large faded cotton blouse was marked with smudges of dirt at the bulges of her sagging breasts, above a full skirt with many pleats which reached almost to her bare feet. And the short, brown, bare stubby toes of those caked and dirty feet reminded me of nothing so much as a row of undersized chocolate candy bars.

On our way back to Salina Cruz late this afternoon we passed a burro plodding along the road. His master was occasionally encouraging him to increase his speed by means of the tap of a small stick. Burrito was piled high with three big bags of some apparently bulky but not too heavy produce. On the very top of the load we were astounded to see tied a small brown pig.

Pig was spread-eagled with his two front hoofs tied together in front, and his two rear ones separated so that each went out to a different side. He had a splendid view of the country as it came into sight, but needless to say he looked quite desolate. While he probably wasn't in any pain, he was undoubtedly cramped and very tired of staying in the same position, although it did save him from trudging miles on his own little hoofs.

It also saved his ingenious master the several extra hours of labor it would otherwise have taken to see that his trudging was done down the road and not into many tempting by-ways he would doubtless have preferred to investigate.

When we arrived at the hotel for dinner this evening, we were anything but happy to see that the *plato del día* was chicken. The all too muscular hens of Latin America commonly earn a precarious living on their own, usually along the roadsides. This necessitates their constant leaping away from burro's and horses' hoofs, and the perilous close rushes of passing cars and trucks. As a result most of them, when finally killed, utterly defy the entrance of the sharpest knives or teeth into their tough, corded flesh. Our experiences with them have invariably left us defeated and hungry. But the hotel here must either raise its own in a chicken yard or, miracle of miracles, know of some white magic to render them tender. For ours this evening was excellent.

April 11, City of Oaxaca, Oaxaca

We started back on the road north this morning, and at noon parked off the side of the highway perhaps a mile away from a large town to have lunch.

While we were eating our sandwiches, we were amazed to see an odd apparition coming down the road toward the town. As it drew nearer, we saw clearly what had made the silhouette so peculiar in the distance.

An Indio boy was pedaling a bicycle which had a large wooden box firmly attached above its rear mudguard. In the box was sitting in solitary splendor a large black ram, not bound in any manner that we could see. Gravely and serenely he regarded the passing scenery, turning his head like a stately male dowager being taken for an airing. And he was still sitting there making no move to escape when they passed from our view, leaving us laughing aloud at so funny and completely unexpected sight.

The sweet and elderly little manager of the Hotel R. just off the central plaza in Oaxaca greeted us with cries of pleasure and surprise when we appeared at the hotel desk late this afternoon. Coming around it to us, he grasped one of my hands in both of his and beaming through his spectacles said with great earnestness, "Oh, Señora, what a miracle it is that you have safely returned from your dangerous trip to the south!"

We smiled a little smugly to ourselves I am afraid, not having been in the slightest danger (aside from that of bad food poisoning in Tehuantepec) and tried to look a little like conquistadores, or possibly Columbus.

He bustled us up to one of the best rooms at the front of the hotel where we can see the plaza by leaning out of the window a little. We bathed, changed, downed two large cold lemonades apiece at our favorite outdoor café, and then completely demolished two excellent dinners in the hotel's dining room. Now we are ready to let the band concert from the plaza, which reaches our room in just the right muted volume, lull us to sleep with the music of some of our favorite composers.

April 12, Oaxaca, Oaxaca

Our field trip today took us out to a group of alfalfa fields down the Valley of Oaxaca south and east of the city. When our work there was finished, we continued on a few miles to the little town of Santa María del Tule. Here, what is said to be the largest cypress or *ahuehuete* tree (*Taxodium mucronatum*) in the world and certainly one of the largest trees of any kind is still growing in the best of health.[2] Cypress trees in Mexico are usually found growing best along the banks of rivers. However, this gigantic specimen is high and dry in a churchyard, and near no visible source of water. It is truly awe-inspiring in its immensity. It is about 160 feet high, and its massive trunk measures 160 feet around at four feet above ground level. To look up from beneath gives one the sensation of being in a lofty forest. Dozens of families of birds can easily nest and live out their years amid only its foliage and branches with plenty of room to spare, as the spread of its enormous branches is 140 feet. According to the information given on a nearby sign, it is estimated that the noble old giant had already been there one thousand years before Columbus discovered America. There seems to be no reason at the present time why it can't continue to live for many, many years more.

Since we were then within a few miles of the famous Zapotec and Mixtec ruins of Mitla, we continued along the road until we came to them. Although the buildings are fewer in number and lack the striking physical surroundings of those of Monte Albán, these are to me much more interesting.

Their architecture is finer and apparently of a higher artistic level. The white stone is sculptured and arranged in a relief in the form of highly intricate decorative mosaic patterns in squares and rectangles along the outsides of the structures. At Monte Albán the buildings are plain, solidly fitted rocks. However, I believe that no finely wrought jewelry of gold or other metal has been found at Mitla. So the talents of the two centers may well have lain along somewhat different lines. Hieroglyphs have been found at both places, those at Mitla being painted in red and black colors inside one building.

April 13, Puebla, Puebla

Our drive today was a pleasant one with no car trouble to take our minds off the enjoyment of the scenery, although Chico still plugs along at a far lower top speed than we think he should and has to be shifted into second or low gear at the slightest grade. We are going to make one more effort to find out in Mexico City what is wrong with his motor.

The mountainous country which we drove through today is beautiful. There are many pine and oak forests on the hills north of Oaxaca. The road follows for miles the narrow "hogbacks" of one winding ridge after another, with steep drops below on either side. Tiny clusters of wattled huts are perched here and there, and their dwellers cultivate the nearly vertical agricultural fields nearby.

We spotted one site where George camped out overnight while attempting to drive from Mexico City via Oaxaca to Guatemala in 1942 . . . an impossibility as it turned out to be after several days on the way. He unhappily discovered, in spite of information to the contrary, that the road at that time ended abruptly at a tremendous unbridged canyon south of Oaxaca. Parts of the old trail-like road are still visible today in places, looking not possibly wide enough for a car to pass along them. They are made up of the most incredibly steep grades, the surfaces of which consist principally of boulders now that maintenance has ended.

The region around Huajuapan is most interesting. Almost every man and boy we passed trudging along the road was weaving a hat of a material that looks as though it might be a grass. It seems odd to have what is usually considered an essentially female activity here taken over entirely by the males of the families. We saw not a single woman weaving, anywhere.

Near Huajuapan the road cuts become, from a geological viewpoint, very different from those through which we had been passing earlier. This area must in the past have been one of great volcanic activity. Volcanic ashes of the most beautiful pastel shades were deposited in thick layers over all of what was then the surface of the earth. They have since consolidated into a fairly soft rock, the colors of which run imperceptibly from pinks and violets to yellows and light greens. In a few places these deposits in the course of road building have already been cut into for thirty-five feet. And they doubtless go considerably below this depth.

An original custom of this region, and of several other widely scattered places I have seen in Mexico, is the means by which the farmers store their corn and other grain harvests. Of course, they could build regular wooden ventilated storage buildings. But that would cost both money and a great deal of labor. Instead, the tops and center of tall candelabrum cacti, or even at times of various trees, are ingeniously cut off. This leaves only a few of the outer stems or branches to act as side rails. Then within these cup-shaped places the crops are packed. They are storage bins that cost nothing more than a little work on the owner's part, and which are completely out of the reach of all the livestock below. I imagine though that both rats and rain take a considerable toll of the amount and value of the stored fodder.

Where small streams crossed the road, there were usually to be seen groups of bare-legged *campesinas,* chattering together as they industriously scrubbed their laundry on the large flat rocks along the banks. In one little grassy valley, literally every spare foot of ground was strewn with drying clothes of every color. It looked as though some ungainly new species of irregular multicolored flower were in bloom.

The vegetation varies greatly from place to place. I have seen a number of plants today for the first time in Mexico. Each one, oddly enough, seemed only to extend within a range of a few miles. One such is a single-stemmed columnar cactus which looks like an underfed skinny saguaro, a cactus familiar to residents of Arizona.

Other clumps of a multi-stemmed cactus are, at this season of the year, thickly studded with two-inch-long, cream, prickly fruits. We saw numbers of birds picking at them and probably using them for a steady food supply. They undoubtedly serve a double purpose. They provide food in the form of seeds and pulp, as well as necessary liquids from their juices, the latter a good substitute for the scarce water in such arid regions.

Driving along south of Puebla, we had an odd sensation. At one point the bed of a small stream was parallel to and a few yards from the highway. High water had so choked it with sand and rocks that it is now actually several feet above the surrounding countryside we were on. Looking up at it gave me the same peculiar tingle up the back that the low country of Holland does, where the land one is on is land only by courtesy of nearby dikes, and the water's surface a hundred yards away is well above one's head.

Once arrived in Puebla this evening, we headed for our favorite restaurant near the central plaza. There we indulged on all the *mole poblano* and enchiladas of various sorts that we could hold. If one must have a vice, there is much to be said for that of overeating. I have heard it staunchly defended in dead earnest by those who should know about such things, as in the long run, the most satisfactory one of them all.

CHAPTER 9

Xochimilco and White-winged Doves in Tamaulipas

April 14, Mexico City, D.F.

We are safely back again after a half day's drive from Puebla. The city is much warmer than it was when we left. Chico has been promptly taken to a garage for a major overhaul. This time we are resolved that he is going to be thoroughly examined and whatever is wrong with him, fixed. George is quite ready to stand over the mechanics eight hours a day if necessary to see that they do what they say they do. In his spare time he has office and paper work which must be done before our next trip. It will be a busy few days for him, though not, this time, for me.

April 16, Mexico City, D.F.

This year we have been purposely waiting all winter until spring and its flowers were surely here to go out and spend a Sunday at the floating gardens of Xochimilco near Mexico City. During the winter tourist season, Sunday is probably the least desirable day to go, as the canals are more than crowded then. But now that Mexico City and its environs are again filled primarily with Mexicans, it is usually family groups that spend their Sundays and holidays at Xochimilco.

After reaching the main plaza by trolley from Mexico City, we spent an hour or so strolling all over the little town and along many of its small secondary waterways that lead to the flower beds of private families. The flowers are lovely, and the kinds are almost innumerable.

It was fun to stop at several little packed-earth landing steps and watch the Indian men decorating their flat-bottomed boats with hundreds of flowers, intricately arranged in complicated patterns. Each boat always has its name painstakingly spelled out in the center of a large flower-covered shield or arch with flowers of contrasting colors.

The boats vary in size from being large enough to carry only two people comfortably, to families of a dozen or more. All of them have four posts with some sort of an awning for shade, several chairs, and a table. The motive power is provided by a man who stands at the back end on a raised lip and slowly poles the boat along.

We finally bargained with a nice-looking lad and hired his craft for a round trip of all of the main canals. It was loudly named Luchita, with bright yellow daisies against a solid background of red carnations. Gradually we moved out from the little side waterway to the nearest main canal and could see a fair amount of traffic already upon it.

Several dozen boats were moving along, about half of them going in our direction. Families of all sizes, and a few couples who were obviously at the courting stage, languidly and happily drifted along under the slight impetus of the push poles. Almost everyone was eating, or drinking beer or cold soft drinks. Each table was piled high with foods of all sorts. Everyone seemed to be happily dipping into them constantly, although it was then only eleven o'clock in the morning. There is an honest reason for so many Latin Americans, both male and female, possessing rotund waistlines by the time they reach thirty.

Dozens of tiny one-man or one-woman canoes frequently plied at a far more rapid rate between the sightseers' boats. Each one was offering something for sale. Many natives had lovely bouquets of pastel flowers, arranged and tied in the center of moist fern leaves.

One sweet-faced young girl paddled up to us with the bottom of her little dugout completely covered with carefully chosen flowers. George bought me, after great deliberation as to choice, a corsage of two red roses encircled by violets and pansies. She made it up as he named the flowers he wanted. The cost was just fifteen cents American, her asking price. Then off she darted with two or three rapid strokes of her paddle to another possible customer, after a "Gracias" and a smile.

We were offered nearly everything imaginable by the floating vendors. The displays varied from fresh heads of lettuce and piles of radishes to barbecued meats, pictures cunningly made of glued birds' feathers or dyed bits of grass stems on wood, all sorts of hand-carved souvenirs, more and more flowers, Coca-Cola, and beer, and to our great amusement, a band.

The latter was complete with five mariachi singers, who were more than anxious to take a tow and ply us with music as we advanced, for a fixed fee of course. We much preferred to drift slowly and quietly along and watch the scenery and our constantly changing neighbors.

The name Xochimilco comes from the Aztec and means "Place Where Flowers Are." For centuries it has been the one place near Mexico City most noted for the flowers it raises. In the early days of the Aztecs many large handwoven reed or cane mats were made and floated on the surface of the lake. Then they were covered with layers of earth. On top of these small huts were built, and flowers and vegetables were planted and cared for.

At first they were light enough to be moved from place to place and were truly floating gardens. But eventually they sank into the shallow water and anchored and rooted there. Slowly through the years more and more earth and humus from the decaying vegetables covered them. Now not only flowers, but tall eucalyptus and olive trees and many shrubs, grow on what seems today to be but patches of solid ground divided by a network of canals.

Everywhere we looked on the land, beautiful flowers bloomed. On almost all of the boats, the señoras and señoritas wore or held bunches of flowers too. It was all very floral and spring-like. We had a wonderful three hours, and when we returned to our starting point, neither of us really wanted to disembark to the everyday solid earth.

When it is seen at its best, a few hours at Xochimilco can still be enchanting. It is an experience that one can have only a few places in the world. But at the height of the tourist season in Mexico it can be a crowded affair, with its charm covered past all discovery by layers of high-pressure salesmanship and noise.

The canals wind so greatly that by boat it can take a long time to move, at the usual slow speed, from Xochimilco Plaza to the public landing, the Embarcadero del Salitre. By land, however, it is only a five minutes' walk. So being very hungry when we finished with our boat, we walked back to the Embarcadero, where there are several restaurants.

Wonderfully enticing odors were seeping out from most of them when we arrived. We picked a small one that seemed already to have almost overflowing patronage, always a good sign, and in it we had a delicious and typical meal.

The pièce de résistance was cabrito, or roast young goat, almost my favorite meat I think, and one which I always ask for whenever I see it on a menu.[1] It was served in the *estilo norteño,* with lots of cooked vegetables, peppers, and rice. We also ordered mounds of mashed and fried black beans, hot tortillas, and coffee with hot boiled milk and had a most satisfactory and filling dinner.

We have now returned to our hotel garden in a semi-somnolent state and are basking in the sun and reading.

April 18, Mexico City, D.F.

At long last after months of speculation we now know that something has been radically wrong with Chico's motor ever since we got him from the factory. It is what caused his continuous and maddeningly slow speeds. Of his four cylinders, tests upon our return showed that he now lacks all compression in one cylinder and about 50 percent compression in a second one. He has been struggling to pull a load of half a ton up and down steep mountain grades on the power of only two and a half cylinders.

To say we are highly disgusted that some of the many garages which have had him in for that purpose couldn't find such a comparatively obvious ailment is to put it very politely. Two of them were supposed to have tested cylinder compression.

We now remember ruefully the many extra hours of hard sitting and pressing a tired foot against the throttle. And we have had to make thousands of double clutching and shiftings into low gear which were really completely unnecessary. At the moment neither of our tempers is at its best. The gloom is very slightly lightened by the knowledge that during the rest of our work here, and most important of all on the long trip back to Denver, Chico will now have more power and speed, although his little four-cylinder engine will continue to be inadequate for use in a truck on mountainous roads.

After another day for repacking of gear, we shall go down to the lagoons near Tampico on the Gulf of Mexico for a week or so of work and study there.

April 20, Valles, San Luis Potosí

We left Mexico City early this morning and by pushing along until a late hour reached Valles this evening. It was an unusually clear day, and our view of the mountains was superb. We stopped for a picnic lunch an hour after going through Jacala, high in the pine-oak forests with hundreds of square miles of mountainous country spread out below and before us.

It was crisp and invigoratingly cool after the heat of the Otomí near-desert country and we enjoyed it thoroughly. Eight long-tailed Hidalgo jays picked just the time we were eating to chase each other in a sparkling display of flashing metallic blues through the tops of some pines directly beneath us.[2]

Shortly afterward while descending the mountain road toward Tamazunchale, we noticed for the first time several very large spider webs strung between the electric light wires along the highway. Six wires covering a width of five or six feet were, in a number of places, completely hung with enormous webs that also extended some ten feet or more along their length. George commented that they looked tough enough to catch small birds. And we could see that they were heavily beaded with entangled insects.

He had hardly made the statement as we drove slowly along looking at one web which hung over an abyss slightly below the road, when we were electrified to see a beautiful little hummingbird darting down straight for it. We waited almost holding our breaths, afraid to see it helplessly caught in the next instant. Instead, it stopped just short of the thick white mesh. It hovered briefly while it deliberately picked out of it with its beak a large fat insect, and then we watched with fascination while twice again it returned and robbed a probably highly irate spider of part of its embalmed hoard.

We saw several pairs of Audubon's [Crested] caracaras, sometimes referred to as "Mexican eagles," today at the lower altitudes. They were perching on dead limbs well above the surrounding vegetation where they had an unrestricted view of things. The handsome birds with their regally held heads, striking black-and-white feather pattern, red faces, and yellow legs, are among the largest members of the hawk family.

As we dropped lower and lower down the mountains, it became hotter and hotter. From Tamazunchale on to Valles we have been in Turkish bath heat.

In order to cool off a bit we went out for a short walk in the late dusk this evening. And we saw a sight that neither one of us has ever seen before. Our attention was first drawn to the magnificent and powerful dives of a taper-winged bird. They extended for a hundred feet or more in the air down almost to the ground level and up again in lightning-fast zooms.

At first I though it was a nighthawk booming. But George soon recognized it as a bat falcon, one of the true hawks. As we stood watching, it dropped down almost to and directly in front of us. Then it turned slightly on its side and in a climbing turn deftly grasped with both talons a small bat, one of a number which had been out flitting erratically after insects for some time.

Its wing strokes slowed until they were barely fast enough to maintain it and its load, and to carry them slightly ahead and upward. We had several minutes of seeing them clearly while it apparently killed its victim by squeezing and digging in its sharp talons. In spite of feeling sorry for the little bat, I couldn't help but admire the dexterity and perfect timing of the maneuver, which had obtained supper for a hungry hawk.

April 21, Tampico, Tamaulipas

The eighty-five-mile drive here from Valles today started out well, for just as we left the town's limits, a white-tailed kite, drifting along in a convection current, started soaring above us in ever-widening circles.

After going through rolling country for a few miles, we entered a narrow gorge with limestone cliffs on either side. Along the cliffs were dotted a number of feather-duster-looking tree-like plants, new to me. Their trunks are superficially of the appearance of old hippopotamus hides. And their basal bulges make them look as though they were suffering from permanent and severe elephantiasis. At the tops of all these bulging trunks are silly little undersized clumps of straggling grass-like leaves, a pitiful anticlimax to the rest of their bulk.

This, George informs me, is scientifically named *Beaucarnea inermis,* which if my memory of high school Latin and French is correct, means something like "unarmed beautiful flesh." The old boy who named it that must have had a sense of humor that one doesn't ever expect and but rarely encounters among taxonomic botanists.

We came out on the edge of an abrupt three-hundred-foot-high limestone escarpment. The coastal plain was directly beneath us and stretched away toward the Gulf of Mexico as far as we could see. Once down, we found the plain very dry, with the only available water a few scattered water holes.

We saw a few white-winged doves and several times heard the raucous chatter of chachalacas calling. We parked off the road for a few minutes so that George could observe the birds, and I noticed a pair of jet, groove-billed anis perched side by side in a nearby small tree. They were industriously billing and preening each other's heads simultaneously, a no mean feat of coordination.

With their oversize and grotesquely hooked, Roman-nose beaks, they always give me the somewhat uncomfortable feeling that they are deliberately and literally looking down their noses at me. It is the same reaction that I have always had to llamas. When living in Peru, I used often to meet llamas along some mountain trail in the high Andes. And I always thought them the most disdainful creatures alive. Their long eyelashes and the manner in which they languidly survey one seem to express utter loftiness and distinct contempt, and a feeling of their being far above the mundane world and its petty two-legged inhabitants.

The scissor-tailed flycatchers must be migrating, as we saw dozens of them today. They are, to me, the most graceful of all flycatchers. Though, perhaps they are not as dexterous as their little vermilion relatives. The scissor-tails have deeply forked tails almost as long as their bodies. As they fly here and there, or two take off together on extended chases and maneuvers, their tails float and flutter along behind them so violently that one can only hope the attachment is a firm one.

We have found a comfortable place to stay out on the beach, or playa, just where the Río Pánuco empties into the Gulf.[3] Tampico, itself, is up the Río Pánuco several miles and stews in very high heat and humidity much of the year. I have been amusing myself this afternoon in a manner of which I am slightly ashamed. By sheer accident I discovered that if, when one of the enormous male blue-black, great-tailed grackles which abound in the neighborhood perched on the porch railing near me, I made a squeaking noise like a wounded bird, he almost always became highly excited. For as long as fifteen minutes at a time he would consistently bristle and fluff to almost twice his size, bobbing his head, lowering his wings, spreading his long tail, and flashing his pale golden eyes. Then in rapid repetition with wide-open mouth he uttered his most penetrating and raucous repertory from squeaks to rusty-gate-hinge squeals to strident whistles.

Over and over the performances have continued as though the grackles were temporarily put under a spell by the alarm notes. Every fiber of their bodies has quivered with intense emotion until it seemed to me that they must drop from sheer exhaustion. But on they went each time until I finally took pity on them and stopped my squeaking. Then after a few minutes of gradually trailing off, they always stopped and flew away.

April 22, Tampico, Tamaulipas

The least terns are beating a steady path past our door, seeking the small minnows that congregate in pools of seawater between the rocks of the nearby breakwater. They look clean and neat in their white coats and black caps. Their dives, as they twist and turn like falling autumn leaves, are graceful precision itself. But their percentage of catches seems low for the number of dives they make. And the fish are so tiny it must take a long day's work for a tern to collect one square meal, even when he is a "least" tern.

A large sinister, triangular fin has just cut an undulating course upstream, going at a good speed. Sinister I still call it because so many travel books read in my childhood associated triangular fins with sharks. And sharks with a taste for lunching off humans.

Now Beebe's studies and those of others seem to prove that sharks are harmless, and that what has been attributed to them is really the unsavory work of the sharp-toothed barracuda.[4] So I suppose I should stop thinking of triangular fins and sinister in the same category. But habit is strong and the feeling still persists, I am afraid.

The gardener's family lives in one corner of the grounds where we are staying. They own a black dog and one rooster, the latter with the longest legs of any fowl I have ever seen. He is mixed reddish-yellow, from a not too distant game-cock ancestor, and an unattractive gray in color. And his habits are to be deplored more than his ancestry.

Both George and I are quite used to having roosters crow in the early morning, and in Latin America where customs among them are apparently different, to their crowing at midnight also. But we are not used to having them deliberately leave their own home and come outside our windows to crow almost all day and at least half of the night at the rate of eight or nine strident crows per hour. This gives one just enough time to fall asleep before being awakened by the next demoniac call.

We feel sure that this rooster has a serious case of insomnia. And we are afraid that soon we will have it too. In our bitter early morning discussions we have found ourselves cataloging all of the horrible methods of forceful deaths we have ever read or heard of, from Caesar's vats of boiling oil to the Sunday tabloids.

But all of them seem to call for equipment which at the moment we lack. Probably in the long run it will be better, if we plan to come back here next winter for further studies, to make the gardener upon our arrival a gift of a dozen hens. That should ensure our sleeping soundly at night for some time to come.

We drove out today to Altamira, the first little town northwest of Tampico. The edge of the big lagoon comes nearby and can be reached by a canal from the town. So we left Chico parked in a pasture surrounded by cows, goats, small boys, and dogs, with coots and jacanas paddling in a pond a few yards away. We put Chiquita in the water of the canal and paddled for half a mile or more along the narrow waterway until it opened out into the lagoon.

Our slow progress along the narrow thread of water, with reeds and sedges rising far above our heads and a complete stillness over everything, was very pleasant. Occasionally a meander would bring us unexpectedly upon a coot family, or a great blue heron poised majestically a few feet in front of us, or a pair of ducks. One and all they hastily pattered away and took wing when they saw us.

The lagoon is so shallow in places that we had considerable difficulty in forcing the canoe through part of it. Probably because of that fact, and because the plant food seems to be scarce, there were very few ducks there. We saw no more than a couple of hundred lesser scaup and coots. Two years ago when George made an aerial survey of the same area, he saw thousands of ducks.[5] However, the water level was very much higher then. To keep us from drifting away from the immediate area we were studying, we paddled over to a large clump of reeds to tie up there. In the center of them on a half-awash log there was perched a yellow-crowned night- heron, head sunk back on his shoulders in slumber. As we softly glided in among the plants, Chiquita's bow gently bumped his log, startling him into wakefulness. He eyed us sleepily for a second or two before he recollected himself and struggled off with a rasping, indignant cry to resume his interrupted nap elsewhere.

April 23, Tampico, Tamaulipas

The great-tailed grackles are making a tremendous commotion this morning. Two males, resplendent in their glossy blue-black coats and waving behind them canoe-shaped tails as long as their bodies, are displaying on two strategically located rocks along the shore.

First one and then the other ruffles his feathers and swells to half again his size. Then he puts his beak straight up to the sky and after prodigious effort and numerous false starts lets out a single strident shriek like the sound of a gate swinging on rusty, disused hinges. This accomplished, he lowers his head and looks triumphantly around to see how many seal-brown lady grackles he has impressed.

Being polygamous, it keeps a male extremely busy impressing his females so that they will not be moved to wander off to some more attractive dark gentleman's bed and board. They are quite fearless. One big fellow has just pigeon-toed his way down the narrow top of a fence some five feet away from me. He swings his hips with every step so that his bouncing tail looks like nothing so much as the beruffled derrière of a *danseuse de can-can.* And he looks me over briefly with his pale amber eyes as he high-steps past.

A large speckled herring gull is flying a patrol along the rocks of the breakwater. He solemnly opens and shuts his beak at regular intervals in a dignified manner as though he were discussing a terribly important subject with himself.

Two little sharp-tailed Inca doves are perched, apparently on the very same pinnacle of a sharp boulder, fifty feet away. Their little shoulders and sides touch as they bask contentedly in the hot sun, dozing or occasionally preening their feathers. They may stay that way for an hour or more if no one disturbs them. The tremendous heat of the midday sun and reflected glare from the sand bother them not at all.

Both the red-headed turkey vulture and the black-necked black vulture are common along the coastal lowlands of eastern Mexico. The black vulture, with its naked, featherless neck and head, is about the ugliest bird I know. But repulsive though it is, it is most useful in all tropical areas where it makes up the only local Department of Sanitation concerned with garbage disposal.

Without numbers of them to rapidly dispose of recently killed animals and other refuse, the resulting stench and epidemics of disease might well drive the people out of some hot coastal sections.

But whatever their worth, their manners are atrocious. They grow so bold and numerous that in some towns in the West Indies and along the northern coast of South America I have actually seen groups of them carrying on open warfare for their food along the gutters and streets with packs of half-starved snarling dogs. And successfully too.

Today a nearby group of five have been fighting among themselves for hours for a few choice morsels. They open their mouths and hiss loudly like snakes as they rush at one another with outspread wings. Although their beaks are large and strong, to my surprise they do not seem to use them at all to peck at each other. One favorite maneuver is to rush at an enemy, head bobbing rapidly up and down, and to turn so that the opponent is butted hard with it. This is very effective when the birds are perched along the ridgepoles or are standing on the roofs of houses, as they often are. It almost always knocks the butted one off balance and makes him fly away. Then the victor gains a few hasty mouthfuls before the vanquished one is able to return and continue the fray.

While we were driving to one of the lagoons for some waterfowl study, I saw what was to me a most unexpected sight. Eight long-billed curlews, each with its sickle-shaped beak curving downward before it, fed in a grassy field, for all the world like a small flock of sheep.

Since they are shorebirds, I have always before seen them wading in shallow water, probing here and there for succulent bits of sea life. It is an odd adaptation for them to take to dry fields too, away from water for feeding and living. But George says they do and that, in fact, the grassy prairie plains of the West and Midwest in the United States and Canada are their natural summer habitat. It is rather late in the season here for them, so we were lucky to see them still gathered in a small group before starting their spring migration back north to breed.

April 24, Tampico, Tamaulipas

We walked out on the mile-long breakwater late yesterday afternoon. A third of the way out and fifty feet away, a large sandbar parallel to it makes a narrow, deeper channel through which the seawater rushes with each wave.

From the top of the breakwater we could look down through the clear water of the channel and see outlined against the light sand schools of fish swimming back and forth.

One large group was made up of perhaps thirty, foot-long mackerel-like fish. They gave us a lesson in precision timing and rhythm since they all flashed their silver sides and bent in perfect unison their deeply forked tails as they swam. Suddenly all was confusion before us.

A three-foot jurel or yellow-tail with wide-open mouth plowed into the middle of the school. It parted, like the burst of a silver signal flare, in all directions. Turning this way and that, the large fish chased the wildly maneuvering smaller ones until all of them had moved beyond the range of our vision.

The lower rocks of the breakwater are apparently hung with translucent green velvet curtains, which upon close examination have turned out to be several kinds of marine algae. As they swing back and forth with the action of the waves, they show rows and rows of tiny black-barred sea snails clinging to the rocks beneath them. Occasionally a crab opens a crack in its fold to peer out and see what edible creature may be within reach of his claw.

Below in the small seawater pools between the rocks are hundreds of young fish. Each one is graded neatly in a group as to size, from quarter-inch kindergarteners to one- or one-and-a-half-inch seniors. They all float up and down and in and out of the trailing seaweed as the waves enter and ebb.

The wind is gusty here along the coast. At intervals it blows the sand with a sharp swishing sound from the high dunes around the sides of the cottage where we are staying. Whenever I hear it, I am immediately shifted back in time to my undergraduate college days at Syracuse University and the wonderful winter nights of skiing on the drumlins there outside the city. The skis and the blowing snow made just that sound as we sped, with awful fascination and the sensation of fantastic speed, down some unknown slope in moon-lit darkness.

The dunes here extend for miles along the coast, making a desolate and excellent hideaway for those who have need of avoiding the attention of the authorities. Some sections are shrub covered, or even support scattered small trees. Others are egg-bald and shift daily or weekly according to the caprice of the winds.

The area near us comes under the latter heading, although here and there the sturdy long stems and thick green leaves of a violet-flowered morning glory are taking root. The plants grow in hit-or-miss, lazy patterns and not only provide a welcome relief from the glare of the pure sand but help bind it in place with their roots. We have found that if the end of a branch with several buds is picked and placed dry on the table or in a vase, one bud will open each day. It will bloom until late afternoon and then close tightly, to open no more. By the next morning another one will have opened and taken its place.

April 25, Tampico, Tamaulipas

We crossed the Río Pánuco today on a small ferry, or *chalán*. Then we drove Chico thirty-five miles south of Tampico to observe the lagoons and marshes in this direction. They had very little birdlife on them, as the waterfowl have already departed for the north. But the migrating birds inland made the trip worthwhile after all. Several flocks of fifty to a hundred white-winged doves, high and to the west of us, steadily followed their predestined ancestral course of migration to northeastern Mexico and Texas.

On one flowering tree close by a small pool of fresh water we saw a dozen kinds of birds. One gave me my best look yet at that marvelous colored little feather bundle, the painted bunting. To a person who has grown up in the northeastern part of the United States where colors are more restrained, a bird which sports a bright blue head, a green back, and a vivid red vest and underdrawers very nearly has to be seen to be believed.

Two little male Maryland [Common] yellowthroats peered at us through black bandit masks as they hopped and posed among the branches. Farther on we glimpsed one beautiful tiny bird, thick-billed and all black except for a bright yellow throat and a golden wash on the wings as though the shaft of each wing feather were yellowish [probably a Yellow-faced Grassquit]. We stalked and stalked it in an effort to collect it. But it always flew on. So now our curiosity has been terrifically aroused. We find ourselves constantly thinking of and discussing the unknown one and wondering what it could have been.

Along the road today we saw three of the toughest-looking cowboys I have ever encountered, driving a herd of fifteen cattle. They made Hollywood's villains look like suave gentlemen's gentlemen.

The fact that the papers have recently been full of news of cattle rustling, that these animals bore three different brands, and that they had the white hides and long droopy ears of expensive imported heat-resisting zebu

White-winged doves over their thorn-scrub nesting brush. The crop in the foreground is sorghum, upon whose seed whitewings feed. In some large nesting colonies such as that depicted, 3–5 million doves raise their young each year. Tragically, such nesting colonies continue to be cleared for agriculture. (George B. Saunders files)

blood did nothing to lessen our suspicions. But in Mexico, not even if you are a Mexican and much less if you are an *extranjero,* do you stop a man on a backcountry road and ask him his business.[6] So we went on, hoping that their good genes would not be wasted in some local slaughterhouse. It is said that many stolen cattle are taken there at once before identification can be made.

April 26, Tampico, Tamaulipas

Fifty feet away from my canvas chair flows the Río Pánuco where it empties into the Gulf of Mexico. Little wind ruffles its smooth surface this morning. The slippery bodies of playing porpoises, looking like wet black rubber, are etched even more sharply than usual against its blue-gray. A dozen or more leap and tumble together, often jumping entirely out of the water. Their tails as they land strike it with the noise of distant cannon fire. And they churn the surrounding area into frothy spume with their exuberant frolicking.

One lone duck wings its solitary way precisely up the middle of the stream. And high overhead a noisy flock of several hundred migrating Franklin's gulls shrilly voices their approval of a strong thermal current within which the birds busily wheel their way upward.

At my feet two digger wasps, each about an inch in length, once again untiringly open and inspect their holes which are dug some three feet apart in the loose sand. The wasps are beautifully marked. Each one has alternating horizontal bands of yellow and black across its long streamlined abdomen, vertical stripes of the same colors on a shield-shaped thorax, and bright yellow legs and head.

How they can tell even the approximate location of their burrows is something I have been completely unable to discover, try though I may. One has just flown in, hovered low over each inch of a foot-square area, and finally settled on a spot within two inches of where I know the hole to be. Walking delicately and pushing the sand behind her with her two front feet, she has tried various places until a giving-away underneath her showed its exact location. Then with fast short strokes of her front feet, she dug the hole out until she could slip down into the slanting tunneled section beneath the sand.

Careful probing by me with a straw has shown this to be from six to eight inches long and slanting forward in direction. Although I have not had the heart to dig it out and see what is there, I suspect a young larva. At least one of her dozen trips a day includes a dead fly clutched between the hind legs below her body. After a minute or two in the tunnel during each visit, out she comes head first. Then she carefully shovels sand behind her until the outer opening of the hole is entirely filled and completely invisible to my eyes.

April 27, Tampico, Tamaulipas

We are staying in a rambling old wooden hotel à la the New Jersey shore resorts circa the early 1900s. There is always a sea breeze, and we invariably enjoy seeing what its rather extensive animal life is up to. Three dogs, two cats, a parrot, and a hen and rooster have the run of the place. And four

The Río Támesi and Tampico lagoons of Tamaulipas as they appeared in January 1947. Although diminished in size, these wetlands still provide wintering habitat for waterfowl. (George B. Saunders files)

white-winged doves, two finches, and two large squirrels live in cages but add much to the interest, and noise, of the establishment. The four doves entrance George. It is always with difficulty that we proceed past their cages into the dining room to eat. Two are in separate cages, and two are in one small cage.

Now that spring has come, it is pathetic, to me, to see the efforts of the two together to nest. No nesting material, or room for a nest, is available. So, one bird sits patiently all day long in the small round clay food bowl on the floor of the cage. The male sings at intervals. When George imitates its call, which he does very well after his years of study of these doves in the field and in captivity, all the males fluff themselves and bob their heads and eye him threateningly with their sharp beaks through the bars. Apparently territorial instincts continue even when the birds have been caged since babyhood.

The whitewings here have been most cooperative with my efforts to write down on paper their territorial songs. Two of the caged birds sing very frequently. So by bringing over my recorder, some lined paper, and a pencil, and sitting patiently at a nearby table, I have now obtained all of the individual songs of these two singing birds.

Whitewing #1 has, besides the usual short "Who cooks for you?" song, three long numbers, each a different variation on the same general theme. Whitewing #2 sings the identical short song. He also has two long ones, each one, however, differing from the other and from the three long ones of Whitewing #1. All are closely related, though, and seem to be variations on the same theme. And the long songs which I recorded in January near Tepic and at the Playa de Cayaco of the whitewings there are all, save one, but slightly different variations. One Tepic whitewing, a western form, and Whitewing #2 here, an eastern form, sang an identical long song, which is very interesting.

The parrot is a charming yellow-headed one named Lorico Rorro. Since he has the run of the place during most of the daylight hours, by means of a little effort and a chocolate cookie or two, he and I have become fast friends. In fact, I have never seen a parrot friendlier with anyone other than its owner.

Lorico positively begs me to scratch his head and stroke his back, and then he certainly gives the impression of beaming with pleasure. Although I suppose that from a strictly scientific viewpoint it is extremely unlikely that parrots can beam. However, he can and does ruffle up all his bright face and head feathers, spread his beautiful tail and dark blue-feathered wings, and trills and whistles his thanks. He will come from a distance when I call him about 95 percent of the time and has developed the desire, apparently, to be our parrot.

Our partings are customarily long drawn-out affairs. George and I are usually followed to the threshold of our front door by him on foot, and by his loud outraged yells as we walk away and leave him.

As we left the hotel restaurant this evening after what was to be our last meal there before returning to Mexico City, a fact which of course Lorico couldn't possibly have known, he startled us all by his unusual actions. We happened to have parked Chico directly in front of the hotel as we sometimes do. As we came out, ready to leave, Lorico was on the second-story porch rail about fifteen feet above us.

As usual I called out, "Adiós, Lorico," as I have a dozen times before with only a spirited verbal response from him. But this time, trembling all over, he launched himself from the railing and flew, clipped wings flapping violently, out and down to where we were parked. He landed on the metal bottom of the canoe which we carry inverted over the jeep. Once there he settled himself on one of the canvas straps which hold it in place and crowed a trill or two in his most triumphant tone.

Then he firmly declined for some time to be removed either by his owner or by anyone else. Needless to say we were highly flattered and greatly touched by such a display of sentiment. If parrots were only a little more adaptable to a life of travel, say, as a dog is, I am sure we would both be doing our best to buy him. But we must continue on without our friend Lorico.

April 28, Valles, San Luis Potosí

Today has been an unusually successful day. Leaving Tampico this morning early, we stopped at a grassy savanna to listen for meadowlarks. It seems to me that we seldom get past any grassy field without listening for meadowlarks if we can possibly spare the time. George's interest in them has been intense ever since he studied them as a graduate student in ornithology at Cornell University.

Today he walked around the savanna with his binoculars, looking for and watching the singing males as they announced their breeding territories to possible interlopers. Since meadowlarks are something that I can take or leave alone, I idly began watching through my binocs a nearby shallow, freshwater lagoon.

Suddenly a line of ten, daintily high-stepping, rose-colored objects began marching across my field of vision. Having seen flamingos living in the highlands of Peru, my first thought was that these were more flamingos. But their necks were not long and curved like snakes, their color was more delicate, and their beaks were long and broad at the end, and yellowy-flesh colored.

It dawned on me all at once that I was, for the first time, seeing roseate spoonbills as something other than an illustration in a *Field Guide to the Birds.*[7] For fifteen minutes or more I watched them as they fed near the shore.

Instead of shoveling ahead of them with their spatulate bills, as I had somehow expected them to do, they swung them like scythes from side to side, directly across their paths, opening and shutting them as they went. Somehow it doesn't seem to me a very efficient method when they could pick up much more at a time by a forward scoop. But undoubtedly they must find that it serves their purpose. Spoonbills have had little spoonbills and continued to feed that way for many years.

I know they are a comparatively rare bird in the United States, being found only in a few colonies in some of the southern states. And their increase or decrease in those is followed each year with great care by ornithologists. So to see ten all at once along the edge of a main highway was tremendous luck. George says there are many more near here and in the Laguna Tamiahua.

It seemed too rare an opportunity not to take the full advantage of. So I put the telephoto lens on our movie camera and slowly walked out through the marshy field toward the lagoon where they were. When I was still a hundred yards away, they started to move nervously and stopped feeding to crane their long necks and look at me.

I started the camera, and it was well I did, as they soon all took off together. But I got a fine colored picture of them as they circled and headed back across the marshes. Doubtless they soon dropped down on a more hidden pond and continued their interrupted meal where they had no candid-camera woman to bother them.

After scouting the lagoon as usual for ducks, we continued along toward Valles. After a time we stopped at a likely-looking spot to see what land bird there might be in the heavy woodland cover. George went one way and I the other to see if there were anything which should be collected.

Altamira, Baltimore, hooded, and orchard orioles were migrating through the trees and shrubs, some of them stopping long enough to sing their liquid, flute-like notes before slipping on. When we came back to Chico again,

George's prize was one of the beautiful little black-and-yellow sprites that we saw and lost near Tampico four days ago.

In the hand it is even more lovely than in the bush. The head, chest, and beak are coal-black, and bright yellow forms a throat patch and little streaks above and below the eyes. But just as an object lesson to show me how easily it is to be absolutely convinced of facts that aren't so when viewing a bird against the light, the rest of it isn't black at all. It is olive-green with a yellowish wash. However, there is no mistaking it for the same bird we saw before.

We know that it is a finch of some kind, but so far have not been able to identify it further as it is apparently a Mexican species and not listed in any of the bird books we have with us. It would be wonderful if it turned out to be a new subspecies. But there is little chance of that it seems. The brightly colored birds are always collected first in a region. To get new forms, George says, get the most nondescript, drably colored, uninteresting (to me) ones there are. And they may be simply fascinating to a taxonomist and prove all sorts of unsuspected things.

Secretly I thought that my best offering was just as fine as his. I brought it out proudly. I had come upon a large black woodpecker tapping its way up the trunk of a tree. It had some white in its wings and a very long, fireman-red crest shaped irregularly like the leaping flame in a fireplace. Under the impression that it was a pileated woodpecker, which I had never seen before in the field, I collected it.

To my astonishment George pronounced it an ivory-billed woodhewer [Pale-billed Woodpecker], and its bill *is* just like white ivory. Knowing that the ivory-billed woodpecker is either recently extinct, or within a few birds of being extinct in the United States, horrible remorse filled me at the thought of having been in any way responsible for lessening its number by one.

My relief was great to learn that this is another smaller species than our extinct one in the States, and that it is locally common within its range in Mexico. We have just made them both into museum skins, and very handsome specimens they are too.

April 30, Valles, San Luis Potosí

The grackles awakened us this morning at five-thirty with their noise. It didn't bother George at all because he had the alarm set for then anyway. We planned to drive out from town and observe the spring migration, which is in full swing. We spent yesterday in the field studying it and saw many large flocks of his beloved white-winged doves, high in the air heading north. So he wants to time the flocks as they go over a given point, estimate their numbers, etc. I, however, had planned to sleep late today, so I was annoyed. I think that of the two, the grackles are worse than the Tampico rooster. One rooster, try as he may, cannot equal the noise produced by thirty or forty male grackles.

Along the front of the cottage where we are staying there is a row of fifty-foot-high royal palm trees in the garden. Their leaves make wonderful landing fields for grackles to crash onto at short intervals. They also make excellent springboards from which to launch themselves, shrieking loudly as they go, to volplane down from beneath our windows.

Directly outside one of our bedroom windows there is a five-and-a-half-foot-high gardenia bush. It now bears fifty-three perfect, waxy-white blooms. I have never smelled such a sweet perfume as fills the room constantly. There will certainly be no need for French *parfum* while we are in Valles.

The ticks are the worst here that we have run into anywhere. Retrieving a shot bird is a feat of willpower. It means balancing between the shrubs and grasses like a ballet dancer to avoid as much as possible their even touching one's legs. In spite of large applications of chigger and tick lotion rubbed all over his coveralls, George got a bad dose of them today. At one time several thousand little seed ticks, each perhaps one quarter of the size of the head of a pin, covered his trouser legs. Most of them could be brushed off, but fifty or so penetrated to skin level.

When he got back to the house, we put all of his clothes into a five-gallon tin. Then we poured a solution of ethylene dichloride with them and closed it tightly to kill the ticks so that he can wear the clothes tomorrow. A careful search has found several already dug in on him. So he will itch from their bites for days to come.

We have recently been told that ticks will not bite one if a big daily dosage of vitamin B_1 has been taken regularly for some time. We have faithfully taken pill after pill of it, but find so far that it makes not the slightest difference to the ticks. They bite merrily on as before, possibly appreciating our higher vitamin content.

This evening at sunset we took a little walk before going in to dinner. There fanned out from a great cumulus cotton-batting cloud in the west a sweeping fanlight of pure gold. It cast sharply angular shadows between its rays of light where the serrated jags of the mountain peaks cut off the light from our plain below. It was striking and beautiful, but violent in its contrasts.

May 1, Jacala, Hidalgo

George had a last early morning spree today observing and collecting. We left Valles about noon to return to Mexico City. We had a very good late lunch at our usual second-story restaurant overlooking the river in Tamazunchale, and then started huffing and puffing up the mountains.

But with Chico's new vim and vigor it has been an entirely different affair from our last December's snail crawl. Where we were in low gear then, today we were in second or even, on the same hills, in high. We made very respectable time for a three-quarter-ton truck.

In May the parched land everywhere is in off-color tints, egg shell, or old ivory, with the most delicate tones of straw or rose-gold added to it. In the mountains the small farmers are once again going through the ancient yearly ritual of burning over their fields.

It is, next to their common practice of cutting down fine old virgin timber on nearly vertical mountain slopes for the purpose of planting a little poorly yielding corn, the most destructive agricultural habit they have. During the day there hangs everywhere a thick pall. The sky is a glassy white with the sun only a faint, diffused deeper cream in it all. As darkness falls, the eager red tongues of the flames can often be seen as they run along the rim of a ridge or curve.

It is beautiful but melancholy to watch. The loss of its trees and watersheds, and the frightful erosion which almost inevitably follows, is probably the greatest enemy Mexico has today. Unfortunately, it is an enemy which is within itself and caused primarily by itself. It is a direct result of the lack of foresighted planning and enforcement to see that its heritage is kept for the good of all, instead of being wasted on the enrichment of a few.

In the late dusk before dinner George and I went out along the terrace of the hotel where we are staying. The lights of the village twinkled on one by one several hundred feet below us in the valley. To the southwest we saw row after row of somber mountains stretch away to the far horizon. A light mist hung over them, veiling everything in robes of gray. High in the heavens the almost new moon glowed yellow.

We were in a calm, peaceful mood when suddenly from above our heads there began a rushing sound, and small black objects hurtled themselves between us and the moon. Squeaking sounds and fluttery flight identified our companions after a few seconds as small bats.

By the hundreds they poured out in a smoky cloud from beneath the rounded rows of red Spanish tiles on the hotel roof. There was a short interval of waiting and then a dozen or more at a time launched themselves from the tiny openings. Each fell perhaps twelve feet until it had attained flying speed. Then up it shot in a climbing curve over our heads.

Intense squeaking was then heard again as the bats which were crowded farther up under the tiles apparently worked their way outward to the opening at the end. Then another group flew out and off. At one time for almost five minutes a terrific squeaking arose and no bats appeared. We could visualize some bats, reluctant perhaps to start the night's work and wishing only for more sleep, blocking the narrow ways down the rows of tiles and causing a traffic jam with their irate, hungry relatives behind speeding them on in no uncertain terms.

By the time total darkness had fallen, which was not more than a quarter of an hour, every bat had gone about its night's occupation of hawking for insects. And we were alone with the moon and the village lights once more. The mountain air seems like wine after the moist heat of the lowlands and we shall sleep soundly tonight.

May 2, Mexico City, D.F.

We left the hotel in Jacala this morning about seven o'clock when its kitchen staff was thinking only vaguely about the possibility of starting cooking sometime. So we breakfasted at a small terraced restaurant down in the village. We noticed when we sat down that an older American couple, obviously tourists and who spoke no Spanish, was at the next table to us. We weren't surprised to find ourselves in conversation with them before long. Chico's unusual appearance intrigues most Americans, especially when it is attached to a U.S. government license plate.

This morning we had parked him by chance directly facing the restaurant so that his one rear license plate and the government shields on the door were not visible from the tables inside. We were surprised at their first question, "Are you a National Geographic Society expedition?" Having read the

National Geographic magazine since earliest childhood and long been fascinated with and yearned to go on such trips as theirs to remote and romantic places, to be mistaken for one in the flesh was quite flattering.

We identified ourselves correctly and explained what we were doing. Several times later during the morning we caught up with and passed their faster car as they stopped somewhere along the way to take pictures. The third time we parked at the same place they had pulled off the road, because I wanted to get botanical specimens of a long-leafed pine I didn't know and which had diagnostic cones.

As we got out of Chico, the woman came over to us and said she hoped we didn't mind their waiting for us periodically and that we'd never know what an affection they had developed for the sight of something representing their own familiar United States government.

Speaking not a word of Spanish, and on their first trip outside the boundaries of the United States, they were in a new car in which they had invested a considerable amount of money. Suddenly they became somewhat overwhelmed at the strangeness of everything and the thought of all the things that could go wrong. We were happy to shepherd them along and recommend our pleasant hotel in Mexico City where we knew they would be comfortable and not overcharged.

We have just met them in the lobby late this afternoon after several hours' separation. They told us again how appreciative of our aid they are and asked us out to dinner. It reminds me rather forcibly of one of my mother's pet axioms when I was a child, and which I was then less than appreciative of: "Cast your bread upon the water and it will return to you."

Now that I am grown, I find it amazing how often it works out just that way.

CHAPTER 10

Hasta la vista, Mexico

May 3, Mexico City, D.F.

WE WILL be here about two weeks this time before leaving on our return trip back to the States. George has a great many meetings planned with both Mexican and American hunters in the city and government officials.[1]

And we must entirely empty Chico, repack in him the things we will need on the homeward trip and in Denver this summer, and store in a safe place those which must stay here for next winter's work. It will be a busy time, and I am not looking forward to it. But like death and taxes it seems to be an inevitable part of a traveling naturalist's life, and that of his wife's too.

May 7, Mexico City, D.F.

Today being Sunday we went down to Sanborn's restaurant for a leisurely breakfast of fresh papaya, hot griddle cakes with maple syrup, and lots of coffee with hot boiled milk. That properly taken care of, we decided to walk over to one of the older sections in the northeast part of the city. The Lagunilla Market there always has a special outdoor Sunday overflow for several blocks along the curbs and streets.

It is impossible to tell from one Sunday to the next what treasures, and what junk, will be on display there. It always varies, and a visit is never, never monotonous. Periodically we go to see whether there is anything we need or can use on display there at the usual low prices. It was as interesting today as always.

First we single-filed our way for a block or two along the narrow remnants of sidewalk left by rows of encroaching small curbstone al fresco restaurants. Wonderful odors constantly arose from the bubbling *cazuelas* of soups and stews, the roasting meats, the frying chalupas, and other things, all of which were being cooked over fragrant charcoal braziers. Then we entered the region of fruits, flowers, and vegetables. The congestion here became so great that to progress at all, we had to take to the middle of the street.

We stopped for a few moments, as always, in the section where the caged birds are for sale and looked to see what ones there were. Whole rows of singing solitaires were simultaneously sliding down their lovely tinkling silvery scales. Mockingbirds, jays, [Northern] cardinals, various thrushes, buntings, and other finches were occupying cages, happily or unhappily according to their moods.

The ones I felt sorriest for were some twenty bewildered cedar waxwings. They were all together in one large cage, frenziedly fluttering against the bars in an effort to escape. Since to the best of my knowledge they never sing in captivity, it seemed a pity that they were there at all.

I must admit, however, that the birds here always seem to have plenty of food and water and are not too crowded. And I have never seen anywhere in Latin America that horrible practice of the Italians—the putting out of birds' eyes with a needle to make them sing more.

Finally, we reached the part of the market which is always the most interesting to me. Here for blocks the curbs and streets are always completely covered with rows of assorted little things, each few feet of space belonging to a separate vendor. There are usually beautiful old French porcelains, English china, Chinese ivory, sterling and plated flatware, old jewelry of every age and fashion, Bausch and Lomb sunglasses, spare parts for cameras, planes, wrenches, Victrola records, books, oil paintings, etc., etc.

Everything is secondhand and in higgledy-piggledy order as far as subject is concerned. But all are neatly spread out for the passerby to view, and it is hoped, to buy. I always feel as though I were moving from the sublime to the ridiculous and back again, to go from some beautiful old, delicate porcelains to a heap of used and bent nails, nuts and bolts, and again to a well-executed oil painting in a wide hand-carved gold frame. One never knows what treasure the next man's little area may include. Which, I suppose, is the real attraction of it all.

After satisfying all our curiosity at Lagunilla, we strolled back past Sanborn's and on in a southwesterly direction to the San Juan Market.[2] This one is noted especially for its fine fresh flowers. All kinds are sold here, either by the single stalk or in pre-arranged designs. We always enjoy walking up and down the streets looking at the hundreds of water jars at the curbs which are filled high with dozens of kinds of flowers.

Today we finally bought some beautiful deep purple gladioli for our hotel room. They have enormous blooms, each one looking as though it had been made out of royal velvet. We also got a corsage of tiny pink rosebuds surrounded by heavenly blue forget-me-nots and green leaves. I didn't particularly need a corsage, but they were so ridiculously cheap and pretty that we couldn't resist them.

We have returned to the hotel to put the flowers in water before they wilt. George is now writing office memos out in the garden in the sun. I am now about to join him with my new reprint of the book by John Lloyd Stephens on his fascinating trips throughout Mexico and Central America in the mid 1800s. During them he rediscovered the major ruins of the Mayas and first brought them to the knowledge of the United States and Europe through his many writings.

May 10, Mexico City, D.F.

Today I decided to go up to Chapultepec Park, which lies to the west of our hotel about half a mile away, and to visit the castle there. On Sundays George and I have several times gone to the park to see the sights. There are always boats on the several small lakes, animals in the zoo, and tremendous and ancient cypress trees. And we love to watch the happy family groups and spirited horsemen on their outings there.

But so far I have never been inside the castle that crowns the top of the very steep hill on the grounds. George has had appointments all day with Mexican hunters and couldn't get back for lunch. So today seemed a good time for me to go.

Today the castle and its immediate grounds are all one great National Museum of History. The majority of its rooms have been left exactly as they were when the emperor Maximilian and Carlota lived in them. To me one of the most interesting *salas* (rooms) there is that in which a magnificent collection of jewelry, principally of the first half of the 1800s and including much of Carlota's, has been arranged on display. The delicate workmanship and design of many of the objects are something one seldom has the opportunity to see today.

The view of the upper terrace is a splendid one if the day is clear of dust storms, as it fortunately was today. Then the entire city of Mexico and its distant surrounding mountains lie spread out before one. I had lunch at a pleasant open-air restaurant in the park under the lofty old cypress trees. Then I sat on a bench until mid-afternoon watching and sketching the few people who were also there.

Just as I strolled back along the Paseo de la Reforma to our hotel, I noticed two bird vendors coming along the street.[3] One carried strapped to his shoulders two parallel rows of little handmade split-bamboo cages. There were fifteen in all, and they towered high above his head some eleven feet or more in the air. In each cage there was a solitaire, or *clarín* as they are called in Mexico. Each one was singing its glorious rushing, tumbling trills under the stimulus of the other birds' songs, the warm afternoon sun, and the slight motion of the cages as the Indio walked slowly along.

His companion carried a single six-foot-long cage, also of bamboo. In it there were almost twenty yellow or red-headed adult parrots [Red-crowned Parrots]. I asked their prices and where they had caught them.

"En la tierra caliente, Señora," replied the men politely.

This is the low, hot country along the Pacific coast. The birds looked well cared for, and each one had its feed dish filled with a cereal mash.

How sweet the *clarín* songs are. I followed the men and their cages at a little distance for several blocks to hear the birds sing. Though I must admit I felt rather like one of the children which the Pied Piper of Hamlin lured on and on with his tunes.

May 14, Mexico City, D.F.

The rains have begun, unfortunately a little earlier than usual. There seems to be no definite schedule by which one may expect them as there is in the

On three occasions, Dorothy remarked about seeing small, melodious, or strikingly plumaged birds being offered for sale in cages by shop or street vendors. Several other sources document that various birds were also being sold for food. This native lad, with his antiquated 32-gauge muzzle-loading shotgun, is a professional hunter. His bag of twenty-four mourning doves was shot on March 20, 1940, near the capital city of Colima in the state of that name. He told George that whitewings, which he also shot, were commonest when the *huizailacaxtle* (a common local tree) and the *chicalote* (prickly poppy) have seeds. Rural-dwelling Mexicans often are remarkably cognizant of the habits of local wildlife. (Source unknown)

West Indies. There one's watch can very nearly be set by the daily afternoon shower.

Here we have awakened to gray rainy mornings. We have also been caught in terrific downpours at noonday. And we have had ourselves drenched by vertical, impenetrable sheets of rain in the afternoon or evening.

They seem to be an elemental force, not at all like the mild patter of our flower-bringing spring showers in the northern part of the United States. Here the clouds slowly mass and remass themselves. Then a pearl-gray suffuses them, all form is lost, and from the hard livid metallic sky the storm starts. The wind flails the trees mercilessly. The thunder of doomsday crashes. And one gets wet, wet, wet if anywhere except under a solid roof.

May 16, Valles, San Luis Potosí

We left Mexico City this morning with Chico packed to the gills, to start our homeward journey to the States. In spite of the beginning of the rainy season, we are lucky in having a fine, clear morning for our departure.

Back we traveled over a by now familiar road. We crossed the dry lake bed, climbed over the high point of the highway, and coasted down into the little irrigated valleys now yellow with standing, ripe, or harvested and stacked wheat.

The glossy black phainopeplas, like sooty cardinals, appeared as always along one dry stretch of the road in Hidalgo where we have never failed to see them.

Through the dry Otomí country there were more beauty and fragrance than we have ever known there in the past. A low prickly acacia shrub, which is usually bare or with dusty gray-green leaves, was having its brief moment of glory and was aglow with delicate powder-pink, fuzzy flowers. Its scent was very strong. At first it seemed saccharine sweet, but suddenly as one sniffed at close range, it turned to a fetid odor.

I saw my first cactus wren here today. It was so large that at first I thought it some sort of a thrush. And a very smart wren he appeared to be. He was slipping along on the ground beneath the pink flowering acacias, watching for the insects which might have been attracted by their extremely strong scent. Upon spying one, straight up into the bush he sprang or flew and seized it. Then he dropped to the ground to eat it, and soon repeated the performance. At the rate he was going while we watched, he soon had a hearty meal.

We stopped for lunch in a small restaurant in Jacala. The open-air dining room was bordered with twelve handmade split-wood cages. In them catbirds [Gray Catbirds], mockingbirds, a solitaire, and a number of other local forms sang much to our enjoyment all the time we were there. Outside, a magnificent, lavender climbing Thunbergiana vine almost covered one end and the roof of the building.

Soon after leaving Jacala, our beautiful weather departed and we were enshrouded in drifting, chilly gray fog. As hills and trees and buildings loomed eerily at us through it, they seemed at first sight anything but the staid objects they really were. Here and there campesinos were steadily weeding their young cornfields, fog or no fog. Both plants and men were clinging to almost vertical slopes in some forest-cleared small patch of rocky soil.

Two solitaires posed for a brief moment on a roadside wire. They were not in the mood for their lovely song, which I would love to have heard once again.

Over several of the dizziest abysses an oropendola [probably a Montezuma Oropendola] had chosen just that spot to hang her foot-and-a-half-long, swinging oriole-like nest from a telephone wire. Her view was superb. And certainly no predator except a winged one could safely reach the spot. But there would be no escape for her young from a violent death a thousand feet below if the gusty mountain winds succeeded in wrenching the delicate fastenings of the nest from its wire.

Down we wound around the mountains, out of the fog, through Tamazunchale and into Valles just at dark. We are comfortably settled in our usual cottage by the palm trees and plan to spend two or three days here while George makes more observations on the white-winged doves and other migrants in the vicinity.

May 17, Valles, San Luis Potosí

George went out early this morning to observe white-winged doves. He came back about eleven o'clock with a long face. The thousands of them we saw each morning and evening when we were here about two weeks ago are no longer to be seen as we expected them to be. They must have migrated on to the north, leaving only a comparatively few resident birds, which will stay and breed here during the summer. These are so scattered in thick, tick-infested vegetation that finding nests and studying nestlings are not going to be practical.

However, George did bring back a fine collection of branches from shrubs which are in fruit now. I have just put them in our plant presses. Neither of us knows the names of the shrubs, so we want to have them identified and then learn the appearance of each one's seeds. Then we can tell in other years from the crops and gizzards of the white-winged doves collected, which of them serve as their food in this region.

In mid-afternoon we both drove back out to the water holes to watch the resident whitewings coming in to drink. At times a dozen or more glided down at once. They looked like a swarm of oversized butterflies because of their striking white bands on their wings and tails, and the odd habit of fluttering their wings back and forth several times as they landed on tiptoe near the water's edge.

With their wings folded, their gray-brown color acts as an almost perfect camouflage against the earth. We could only pick them out clearly then by using our binoculars. They drank daintily and briefly, for a few moments only. Then up and away they darted in powerful, driving, zigzag flights.

As we sat there miles from any human habitation, George happened to look in the rearview mirror of the jeep. He told me to look quickly at the road two hundred yards behind us. There was crossing in a most leisurely manner a fine large wild turkey hen. Chestnut wing, tail, and rump feathers showed her clearly to be a wild one. And we know that they are still found in this general area in fairly large numbers.

Scarcely had she disappeared before I saw, motionless on the opposite side of the road from us, my first tinamou [Thicket Tinamou]. It observed us quietly for a moment or two before fading away into the undergrowth. We have heard their rather sweet, plaintive whistles not far away from us each time we have been in this vicinity. But today is the first time I have caught sight of this odd, terrestrial bird.

The greatest moment of the day, however, happened on our way to Valles. Driving along very slowly, we looked down by chance at a roadside pool, and there were two black-bellied tree ducks [Black-bellied Whistling Ducks] feeding in the shallow water. Standing high on their pink legs, they looked almost goose-like to me.

Quickly the 12-gauge shotgun was gotten into place, since we know that the museum was anxious to have another study skin of them. And I turned the nearer one into a very dead duck indeed.

Shortly afterward we saw five more of them flying over the road at treetop level, but out of range. Then for the first time I had a chance to see their beautifully marked black-and-white wings in flight. They are most unduck-like in some of their habits. They actually nest in hollow trees, from where they get their name of course.

Shortly before dinner this evening we were sitting out in front of our cottage on the lawn. As we chatted, George happened to look over at the fence which separated us from the Pan-American Highway. There, perched on the post directly in front of us and only a few feet away, was a solemn little owl. Straight as though he had a miniature ramrod for a spine he sat, regarding us steadily. He was a ferruginous pygmy-owl, the first I have ever seen, and so a new entry for my life list. And pygmy he certainly is. It is for once a most appropriate common name.

He and we sat companionably there together for half an hour or so until hunger drove us on to dinner. When we returned, he was gone, although I went out at once to see if by chance he were still there.

May 18, Ciudad Mante, Tamaulipas

We packed up this morning and drove slowly along, observing birds and plants as we went. We arrived here just before noon. It has been a scorchingly hot day, over 100°F in the shade. As there have been no rains yet this year in this region, everything is burned to an unattractive brown crisp.

We spotted six chachalacas this morning. Three of them were in the same tree about twenty-five feet from the road, seemingly not at all wild. It surprised me to see them so tame and still enjoying good health. They are pheasant size and excellent eating. And the Indios always try to catch them for food when they make an appearance.

Their onomatopoeic name comes from their call, which is usually repeated over and over again at the top of their grating, raucous voices. They have an original method of singing which can most nearly be compared with the old English "round." A male starts his song of *cha-cha-lac, cha-cha-lac, cha-cha-lac,* etc., in a low key. Then his female starts hers, being always just one syllable behind him and in a considerably higher key. On and on they go in

Cerro de Bernal, Tamaulipas. Millions of white-winged doves still nest along the brush thickets bordering the Río Guayalejo in the foreground. (Photo by Pierce B. Uzzell)

unison, two or three of them sounding like an international convention once they really get started.

And they start at the slightest provocation, such as sunrise, or sunset, or any unspecified hour in between. I speak strongly about them since the night we inadvertently made camp after dark apparently in the middle of a chachalaca settlement. We were awakened before dawn and kept awake thereafter by their "singing."

Shortly before we reached Ciudad Mante, we began to see flock after flock of migrating white-winged doves crossing the road, headed north, high above us. In twenty to thirty minutes George estimated that several thousand birds had flown before us. They were divided into strung-out flocks of from about twenty-five to seventy-five individuals each, in typical migration formation.

As they flew along at high speed, it was a thrilling sight to see them. And to realize the hundreds of miles they had already covered and the urge that was forcing them on at their best pace. We are going to spend an extra day or two here so that George can do more observing.

This afternoon we drove out along the Mante-Tampico road and spent several hours looking for and listening to nesting whitewings. As I was sitting quietly in the car while George was off in the woodland, a male red-eyed cowbird [Bronzed Cowbird] put on his courting flight and display not fifty feet away from me. First while on the ground, he fluttered his wings violently for a moment or two so that he appeared to vibrate all over. Then he raised his neck ruff, fluffed his feathers, and gleamed his brilliant red eyes at the female for all he was worth.

Suddenly he rose straight up in the air, and there he hung for almost two minutes like a helicopter, hovering a yard or so above his lady fair. She, as far as I could judge, paid little or no attention to his efforts. Then he dropped to the ground and regarded her ardently, whereupon she flew off in a most indifferent manner. The last I saw of them he was pursuing her with all the perseverance that a true gallant should show.

A little later while I was still sitting in Chico, half a dozen cattle came out of the underbrush and to a nearby roadside pool of water to drink. Then they straggled on as the mood struck them, past the side of the jeep and down the road. A roadrunner, *paisano* or *Correcaminos* in Spanish, accompanied them. He was apparently under the impression that if they showed no alarm, there was no need for him to either.

So as I sat motionless, he spent ten minutes or so not more than twenty-five feet away from the car. He darted after lizards at times in short, sudden rushes. But mostly he simply alternately raised and lowered his tail and crest in a jerky but exceedingly nervous manner. His mannerisms were not only amusing but strongly tinged with vanity. The silly little dark topknot on his paler head reminded me very much of a feather duster.

May 21, Monterrey, Nuevo León

We spent the past two days on more white-winged dove studies near Ciudad Mante, and left there this morning. We were rather stolidly driving along with little to relieve the monotony of the grayish-brown scenery when we came over the rise of a little hill. There directly in the middle of the road we saw a plump, bluish-gray bird standing. It was drawn up to its greatest height as if to watch us better. As we bore down upon it, I clearly saw a buffy topknot raised on its head before it ran at top speed across the road and into the brush. And so I added the first scaled quail or "cotton top" to my life list.

About forty miles north of Ciudad Mante we turned off onto a side dirt road and drove along it for several miles to a rancho where we knew that several other biologists from the States were working and studying. Six graduate students from the Universities of Michigan, Cornell, Minnesota, and Penn State are there, collecting various birds and small mammals from the general area. We had a pleasant but all too brief chat with them, made inquiries about mutual professors and friends, and continued along our way.

For twenty miles or more we passed among very extensive fields of citrus fruits, corn, and assorted vegetable crops. Occasionally we could glimpse white-winged doves migrating along parallel to us in a northerly direction. High ranges of the Sierra Madre Oriental, covered with heavy clouds that extended out over the plains to us and gave some relief from the intense heat, rose to the left of our road. Then the country changed quite drastically as we neared the Mesa de Llera and became dry and much more restricted in vegetation.

The mesa is, from a geological viewpoint, a very interesting spot. We climbed the road up along a ridge that seems to be one side of an enormous round crater, which has a span of several miles across its widest part. In the bottom of the huge bowl, probably two miles apart, are two volcanic cones. One is so eroded that only the hard central core remains above the ground level. It must have been a tremendous and rather terrifying affair before erosion took its customary toll.

We stopped again at Montemorelos, where we bought the very delicious tangerines last December, to buy some more. We were highly disappointed when we peeled them to find that every one had dry pithy tissue, completely without juice. Last winter they had an unusually severe "norther" with temperatures below freezing all through the region, which killed many citrus trees and other crops. So the freeze must have partially frozen the fruits which did survive, and they did not develop as they should have.

There is one bird which we have seen constantly all through our stay in Mexico, from Texas to Chiapas, and both in the highlands and the lowlands. In fact, there have been few days when we have missed seeing every few miles or so, sprucely and erectly sitting on some wayside pole or branch, the trim shape of a sparrow hawk [now American Kestrel]. It has finally become almost a game with us to see which will first spot the next one, "Samuel" we call them all collectively. Invariably the next one sits as though he had merely flown on ahead and was patiently waiting for us slowpokes to catch up with him. The males are especially beautiful with their rust-red crowns, back, and tail; slate-blue wings; sharp eyes; and deeply curved beak.

Today Samuel has outdone himself and has greeted us, it seems, from every other telephone pole and dead tree. Sparrow hawk is really a very inaccurate name for him, falcon that he is. Grasshopper hawk would be one much more to the point since the bulk of his food is grasshoppers, varied by meadow mice. It is a comparatively rare little bird that feels his talons.

But in the early days of American natural history, the visiting European naturalist who first saw, described, and named him, was confused and thought him identical with the European sparrow hawk. The European one is a form which, like our sharp-shinned hawk, does eat small birds in quantity. Hence the unearned blot upon what is in truth a practically pure escutcheon.

We drove for many miles through country where the plant life is made up mostly of white flowering cordias, mesquites, Texas ebonies, huisaches, cat-claw and bull-thorn acacias, and tall columnar cacti. Finally, near Linares we again came into extensive citrus groves. And they were pretty much on the scene until we arrived in Monterrey.

The mountains around Monterrey are as lovely as ever. This evening they have been especially spectacular. Trailing gray clouds have covered the highest jagged peaks, and intermittent thunder and lightning have flashed and rumbled for an hour from a ferocious storm which is going on around their heights. We are sitting in our deck chairs in front of our cottage with a cool breeze running down from the mountains. We have been watching the spectacle in dryness and comfort, the proper way to enjoy a severe thunderstorm.

May 22, McAllen, Texas, U.S.A.

The fog was hanging heavily over the mountains as we started out from Monterrey this morning. We were very happy to see it, as it cooled for a little while what later became blast furnace heat.

The drive from Monterrey to Reynosa, the town on the border directly across from McAllen, Texas, is just about as uninteresting as any part of Mexico I have seen. The land slopes down very gradually to the dry coastal plain, which is covered mostly with a low cactus-yucca-mesquite association of xerophytic vegetation. It has a dusty gray-green sameness in color which is relieved by nothing except an occasional small herd of goats or scattered dismal-looking run-down ranches or farm houses. And it is hot, hot, hot!

Several rabbits were startled at our passage today. One large jack bounded off across a field in tremendous leaps just after we had gone through the town intriguingly named General Bravo, for General Manuel Bravo.

The only birds we saw were a spiral of white-necked ravens in a convection current, one group of twenty northward migrating Franklin's gulls, and the usual mockingbirds, roadrunners, and ground, Inca, and mourning doves. Although we expected to see our migrating whitewings again, even they failed us. I couldn't blame them. I don't see why any animal would voluntarily live or cross there at this season so long as there were anywhere else in the world to go.

We reached the border at twelve-thirty this afternoon. In an hour or so we were through both the Mexican and the United States customs, and once again on native soil. We headed immediately for a favorite restaurant and had all the things we've been longing for, for months, but which because of one disease or another we thought it wiser not to eat. My lunch was a highly satisfactory mixture of ice cream, iced coffee, lettuce, raw sliced tomatoes, butter, and a milk shake, and I am still replete but happy.

Just five months ago today we left McAllen for the beginning of our Mexican winter. We have had a wonderful trip, have seen many places and people, and have learned many things we didn't know then. The work has gone well, and neither George nor I has had any serious illnesses. All in all it has been another fine time in our lives.

It is always fun to go, but fun to be back home too. Our Mexican trip is formally ended for this year. Now we have remaining the drive back to Denver, the unpacking and sorting out of our load, and weeks and months of work on the identification and study of the birds and plants we have collected. Then too, there are various reports to be written. So it is *hasta la vista* for Mexico from all three of us (Chico joining us) for this year.

Epilogue

by Jesús G. Franco Pizaña

MANY A thing has changed in Mexico since the time Dorothy and George Saunders traveled across the country nearly sixty years ago. Many of the dusty, suspension-breaking roads Chico painfully crawled through are now modern, multi-lane, public and toll roads that save the traveler considerable driving time and grief. The joyful rides aboard man-powered *chalanes* Dorothy so much enjoyed have been replaced with fast commutes across sleek bridges. Many of the sleepy, picturesque, little towns they visited and effortlessly criss-crossed on foot are now bustling urban centers and modern tourist resorts. Very few locations off the beaten path still remain in what Dorothy called a "lovely rare state of being completely non-tourist conscious."

The elusive, brush-guarded Laguna de Tres Palos near Acapulco that the Saunderses had a hard time finding access to is now completely surrounded by roads. The distant mountains surrounding Mexico City sixty years ago seem not as distant today. Instead, they are smothered in both regulated and unregulated housing developments precariously clinging to their sides and now are part of the ever-growing urban landscape of one of the largest cities in the world. White-winged doves, as if following in the steps of many of their human counterparts, have all but abandoned their rural homes in favor of the ever luring modern urban landscape. This has raised concerns over the long term consequences of the phenomenon, puzzling biologists who try to explain the driving forces behind it. Today's vastly improved communications make the modern traveler wonder how anybody could endure, like the Saunderses did, five months of relative isolation from the rest of the world without going insane.

Over the years, winter waterfowl surveys have continued to be conducted in Mexico by U.S. Fish and Wildlife Service biologists. Such surveys have served as an extension of the traditional Midwinter Waterfowl Inventory conducted every year in the United States and parts of Canada. The general objectives of the Mexican surveys are to monitor winter habitat conditions and utilization by waterfowl as well as the birds' distribution patterns and population trends (Thorpe et al. 2006).

Observations made by waterfowl biologists conclude that the east coast of Mexico, with its vast and complex systems of wetlands, continues to be an extremely important and valuable resource for waterfowl as well as for many other species of birds. Marshes and coastal ponds in the Laguna Madre and Tamaulipas Lagoons support Redheads and Northern Pintails whereas marshes and natural wetlands in the Rio Grande and Pánuco River deltas, with their extensive submersed aquatic vegetation expanses, support Northern Shovelers, Lesser Scaup, Gadwall, American Wigeon, Northern Pintails, American Coots and many others. Other localities along the Gulf of Mexico coast offer an extensive array of waterfowl havens ranging from coastal lagoons and mangroves to freshwater marshes, interior flooded agricultural areas, and rice fields, all providing habitat for other bird species such us Black-necked Stilts, Northern Jacanas, Wood Storks, Roseate Spoonbills, various egrets and herons, and many more (Roetker and Wortham, 2006).

However, waterfowl biologists have also observed how scores of important ephemeral ponds in the Rio Grande delta have disappeared due to agriculture and other land use practices, and how the increased disturbance of otherwise remote, tranquil wintering areas, by fisherman, speeding boats, and even hunters is altering feeding and loafing behavior of birds, influencing their numbers.

Winter waterfowl surveys also have continued to be conducted on a triennial basis in the Interior Highlands and the Lower West Coast of Mexico. Although comparability between survey years and long-term averages can be difficult to define at times, recent total population estimates of ducks and Lesser and Greater White-fronted Geese appear to be higher than in previous surveys. Estimates for non-waterfowl species like the American Coot and Sandhill Crane also appear to be higher. However, recent weather events in the form of severe to extreme drought also have exacerbated wetland condition deterioration in areas of Oaxaca, Guerrero, Jalisco, and Michoacán. Additionally, waterfowl biologists have indicated that important freshwater marshes might soon be drained and need to be protected. Despite the apparent population estimate increase observed during recent surveys, an analysis of the long-term trend is leading biologists to believe that a decline of the overall populations of ducks is real (Thorpe et.al. 2006).

But also, many things are yet to change in Mexico. In spite of significantly improved and streamlined procedures, modern border crossings into Mexico can be as painful as they were when the Saunderses crossed the border in 1949. Long lines, multiple copies of every imaginable paper under the sky, downed computer systems, and undertrained, overworked and underpaid customs officials create a perfect environment to challenge and reveal the limits of one's true patience.

Despite federal government efforts to regulate the capture and commercialization of wild song birds, the birds continue to be sold indiscriminately not only in domestic markets, a common practice since pre-hispanic times, but now also in European and Asian markets. Tradition, poverty, social activism, corruption, and trading mafias all combine in a convoluted play in which only a few greedy "bird lords" win by hoarding the lucrative profits of the activity. The vast majority—the bird trappers, merchants and peddlers—come out even by making just enough to make it through the day, and the song birds always lose.

Personal and candid conversations with top ranking Mexican wildlife officials have revealed their full awareness of the problem as well as their utmost frustration for not having the resources and law enforcement support necessary to do much of anything about it. Paradoxically, one of the largest threats to wild birds worldwide in recent years, the avian influenza or bird flu, is helping to keep some of the wild birds wild. Fears of triggering a worldwide bird flu pandemia has resulted in a reduced demand for song birds in international markets.

Environmental degradation is still rampant. Strong human population growth is one of the main forces driving wildlife habitat loss due to wetlands drainage, air and water pollution, rangeland overgrazing, and forest clearing. Deforestation, watershed depletion, and subsequent erosion continue to be some of the greatest threats to Mexico's environment today as a result of the seemingly perennial lack of planning and enforcement. Just like in 1949, the environment in Mexico continues to be ransacked for the economic benefit of only a few.

For more than half of Mexico's forty million poor, subsistence farming, livestock production, and hunting, usually associated with vegetation clearing and unregulated hunting and fishing, are the only viable means to put food on the table. Areas of southern Mexico with some of the highest biodiversity are also home to some of Mexico's poorest people. Environmental degradation is also associated with wealthier landowners and farmers. The use of modern irrigation and farming practices in intensive agriculture systems has resulted in salinization, soil erosion, soil and water pollution, and desertification (Valdez et al. 2006). Decades, perhaps centuries of either greed or poverty, most of the time both, have engendered an increasingly imperiled relationship between the people and the land. This has resulted in the raping of its natural resources, effectively perpetuating poverty across most of rural Mexico, which fuels social unrest and a steady, hope-filled exodus for the northern border.

Many of Mexico's environmental problems are the direct result of the government's failure to integrate sound, long-term, national development and conservation policies. The hugely overwhelming economic and social problems, recurrent budget deficits, land tenure systems, and national debt have prevented federal and state governments from adequately funding natural resource management and conservation programs. Simply stated, wildlife is not a national priority. Lack of adequate funding, high employee turnover, and the continuous shifting of government agencies has prevented the establishment of a robust wildlife program in Mexico (Valdez et al. 2006).

Nevertheless, as wildlife is considered an increasingly important, economically viable resource, more and more conservation and management efforts are being conducted by different stakeholders. The recognition of the importance of wildlife conservation has resulted in a recent speedy increase in the creation of federal programs and laws to administrate and manage wildlife populations.

Much legislation to regulate hunting and protect game and non-game species in Mexico has come and gone. In 2000, a landmark wildlife conservation and management law titled General Wildlife Law was issued. Touted as one of the most comprehensive wildlife legislations ever enacted in Mexico, the law contains a wide array of provisions for the sustainable use of wildlife. Some of these provisions include incentives for landowners; increased cooperative work among federal, state, and private individuals; establishment of priority species and their habitats; development of wildlife research centers; use of wildlife by indigenous people; development of environmental education programs; law enforcement training; and the establishment of restrictions on exotic species (Secretaría de Medio Ambiente y Recursos Naturales, 2002).

The federally centralized responsibilities for the conservation and management of wildlife in the entire country have been delegated to a number of state governments as part of a much needed process of decentralization. The U.S. Fish and Wildlife Service and several wildlife state agencies, notably the Texas Parks and Wildlife Department, Arizona Game and Fish, and New Mexico Game and Fish have played key roles in assisting both the Mexican federal wildlife agency and the newly created Mexican state wildlife agencies get off the ground as fast as their limited economic, professional, and human resources allow it.

In the context of Mexico's highly complex nature, composed of a rich collection of megadiverse ecosystems, cultures, socioeconomic levels, and land tenure systems, Valdez et al. 2006 offered a series of strategies for resolving the country's most pressing wildlife management challenges. Policies need to include localized and strategic approaches and concerted efforts to address each of the many aspects of the multifaceted conservation challenges. Since neither federal nor state governments are able to adequately fund, staff, and administer viable wildlife programs, partnerships with the private and corporate sectors are highly encouraged. Ideally, the integration of priority species programs should consider economic incentives programs that economically benefit landowners, wildlife, and wildlife habitats.

Partnerships with and between nongovernmental organizations and academic institutions can facilitate the establishment of long-term research and monitoring programs generating, among other things, critically needed data throughout the country on species distribution, status, demographics, and habitat quantification. A crucial element in improving wildlife conservation in Mexico is the establishment of undergraduate wildlife academic programs to develop urgently needed wildlife biologists. Cooperative wildlife programs between American and Mexican universities can encourage the recruitment of Mexican wildlife students, the development of collaborative programs, and joint projects. Wildlife knows no borders, therefore it is of utmost importance that both American and Mexican wildlife biologists be equally educated, trained and informed to best protect wildlife and wildlife habitat.

Many of these strategies are currently being addressed through the binational borderlands programs of U.S. state wildlife agencies like the Texas Parks and Wildlife Department, Arizona Game and Fish, and New Mexico Game and Fish. Through direct, one-on-one collaboration, U.S. and Mexican border states are working on developing binational, bilingual, and bicultural wildlife conservation programs focusing on biodiversity and sustainable development.

Through an array of programs like Wildlife without Borders, Partners for Wildlife, and Joint Venture projects to name just a few, the U.S. Fish and Wildlife Service and its Mexican counterpart are actively engaged in promoting sound, science-based wildlife research, management, and conservation initiatives. Strong partnerships, trust, and a clear understanding of the cultural and socioeconomic differences between the partners is crucial in the successful implementation of these initiatives.

Collaborative work currently being conducted addresses binational issues, needs, and priorities. Capacity building that promotes professional and institutional development, wildlife resource administration, technical guidance, big game, upland and migratory game bird—most notably White-winged Doves—and waterfowl conservation and management, development of wildlife corridors, environmental education, ecotourism development, and hunter education program development are only a sample of the wide range of priorities currently being tackled.

Mexico's biodiversity is top ranking in the world, and it serves as head of a wildlife dispersal corridor. As such, the country's commitment to actively participate in international forums like the Trilateral Committee for Wildlife and Ecosystem Conservation and Management, the Border Governors Conference, and many others, has played a most essential role in helping prioritize, integrate and advance Mexico's conservation agenda for the benefit of the wildlife of all of North America.

For a country where wildlife is not a priority, the commitment by wildlife government agencies, nongovernmental organizations, landowners, corporations, and private individuals to address the urgent need to resolve their seemingly insurmountable wildlife conservation challenges is something to be applauded. Without a doubt, the actions and management decisions they make today will have long lasting effects, either positive or negative, and most certainly will shape the state of their natural world for present and future generations.

Appendix: Bird List

Name Given by Dorothy	Scientific Name	Present Common Name	Mexican Name
anhinga	*Anhinga anhinga*	Anhinga	Ahuizote
ani, groove-billed	*Crotophaga sulcirostris*	Groove-billed Ani	Garrapatero
ani, smooth-billed	*Crotophaga ani*	Smooth-billed Ani	Pijui
avocet	*Recurvirostra americana*	American Avocet	Picocurvo
baldpate	*Anas americana*	American Wigeon	Pato chalcuán
bird, thick-billed, black and yellow	*Tiaris olivaceus?*	Yellow-faced Grassquit	Zacatero oliváceo
bittern, American	*Botaurus lentiginosus*	American Bittern	Torcomón
blackbird, Brewer's	*Euphagus cyanocephalus*	Brewer's Blackbird	Tordo de ojos amarillos
blackbird, yellow-headed	*Xanthocephalus xanthocephalus*	Yellow-headed Blackbird	Tordo cabeza amarilla
booby, brown	*Sula leucogaster*	Brown Booby	Bubia vientre blanco
bunting, indigo	*Passerina cyanea*	Indigo Bunting	Azulito
bunting, painted	*Passerina ciris*	Painted Bunting	Sietecolores
canvasback	*Aythya valisineria*	Canvasback	Pato coacoxtle
caracara	*Caracara cheriway*	Crested Caracara	Quebrantahuesos
cardinal	*Cardinalis cardinalis*	Northern Cardinal	Cardenal común
catbird	*Dumetella carolinensis*	Gray Catbird	Pájaro gato ceniciento
chachalaca	*Ortalis leucogastra*	White-bellied Chachalaca	Chachalaca vientre blanco
chat	*Geothlypis poliocephala?*	Gray-crowned Yellowthroat?	Verdín carbonero
coot	*Fulica americana*	American Coot	Gallareta gris
cowbird, red-eyed	*Molothrus aeneus*	Bronzed Cowbird	Tordo ojirojo
crow, Mexican	*Corvus imparatus*	Tamaulipas Crow	Cacalote tamaulipeco

Name Given by Dorothy	Scientific Name	Present Common Name	Mexican Name
cuckoo, rusty ground	*Morococcyx erythropygus*	Lesser Ground-Cuckoo	Cuclillo chiflador
cuckoo, squirrel	*Piaya cayana*	Squirrel Cuckoo	Vaquero
curlew, long-billed	*Numenius americanus*	Long-billed Curlew	Zarapico
dickcissel	*Spiza americana*	Dickcissel	Gorrión cuadrillero
Dives dives	*Dives dives*	Melodious Blackbird	Tordo cantor
dove, Inca	*Columbina inca*	Inca Dove	Tortolita común
dove, Mexican or ground	*Columbina passerina*	Common Ground-Dove	Tortolita
dove, mourning	*Zenaida macroura*	Mourning Dove	Huilota
dove, white-fronted	*Leptotila verreauxi*	White-tipped Dove	Paloma suelera
dove, white-winged	*Zenaida asiatica*	White-winged Dove	Paloma de alas blancas
dowitcher	*Limnodromus scolopaceus*	Long-billed Dowitcher	Agachona piquilargo
duck, black-bellied tree	*Dendrocygna autumnalis*	Black-bellied Whistling-Duck	Pichichí
duck, fulvous tree	*Dendrocygna bicolor*	Fulvous Whistling-Duck	Chiquiote
duck, Mexican	*Anas platyrhynchos diazi*	Mexican Mallard	Pato triguero
duck, ring-necked	*Aythya collaris*	Ring-necked Duck	Pato chaparro
duck, ruddy	*Oxyura jamaicensis*	Ruddy Duck	Pato tepalcate
egret, American	*Ardea alba*	Great Egret	Garzón blanco
egret, snowy	*Egretta thula*	Snowy Egret	Garza nívea
Empidonax	*Empidonax* spp.	Unknown species of *Empidonax*	Mosquero
falcon, bat	*Falco rufigularis*	Bat Falcon	Halcón garganta blanca
finch, house (linnet)	*Carpodacus mexicanus*	House Finch	Gorrión común
flycatcher, derby or kiskadee	*Pitangus sulphuratus*	Great Kiskadee	Luís Grande
flycatcher, scissor-tailed	*Tyrannus forficatus*	Scissor-tailed Flycatcher	Papamoscas tijereta
flycatcher, silky	*Ptilogonys cinereus*	Gray Silky-flycatcher	Capulinero gris
flycatcher, vermilion	*Pyrocephalus rubinus*	Vermilion Flycatcher	Cardenalito
gadwall	*Anas strepera*	Gadwall	Pato pinto

gallinule, purple	*Porphyrio martinica*	Purple Gallinule	Gallareta morada
gnatcatcher, blue-gray	*Polioptila caerulea*	Blue-gray Gnatcatcher	Perlita común
goldfinch	*Carduelis tristis*	American Goldfinch	Dominiquito viajero
goldfinch, Mexican	*Carduelis psaltria*	Lesser Goldfinch	Dominico
goose, white-fronted	*Anser albifrons*	Greater White-fronted Goose	Oca salvaje
goose, snow	*Chen caerulescens*	Snow Goose	Ansar blanco
goshawk, Mexican	*Accipiter gentilis*	Northern Goshawk	Gavilán pollero
grackle, great-tailed	*Quiscalus mexicanus*	Great-tailed Grackle	Zanate
grebe, Mexican	*Tachybaptus dominicus*	Least Grebe	Zambullidor chico
grebe, pied-billed	*Podilymbus podiceps*	Pied-billed Grebe	Zambullidor pico pinto
grebe, western	*Aechmophorus occidentalis*	Western Grebe	Achichilique
grosbeak, black-headed	*Pheucticus melanocephalus*	Black-headed Grosbeak	Tigrillo
grosbeak, blue	*Passerina caerulea*	Blue Grosbeak	Picogordo azul
gull, California	*Larus californicus*	California Gull	Gaviota califoriana
gull, Franklin's	*Larus pipixcan*	Franklin's Gull	Apipizca de Franklin
gull, herring	*Larus argentatus*	Herring Gull	Gaviota plateada
hawk, black and white	*Herpetotheres cachinnans*	Laughing Falcon	Guaco
hawk, Mexican black	*Buteogallus anthracinus*	Common Black-Hawk	Aguililla cangrejera
hawk, Sennett's white-tailed	*Buteo albicaudatus*	White-tailed Hawk	Aguililla coliblanca
hawk, sparrow	*Falco sparverius*	American Kestrel	Cernícalo chitero
heron, black-crowned night	*Nycticorax nycticorax*	Black-crowned Night-Heron	Perro de agua
heron, boat-billed	*Cochlearius cochlearius*	Boat-billed Heron	Macaco
heron, great blue	*Ardea herodias*	Great Blue Heron	Garza morena
heron, little blue	*Egretta caerulea*	Little Blue Heron	Garza azul
heron, Louisiana	*Egretta tricolor*	Tricolored Heron	Garza flaca
heron, yellow-crowned night	*Nyctanassa violacea*	Yellow-crowned Night-Heron	Pedrete enmascarado
hummingbirds	?	Hummingbird	Chupaflor

Name Given by Dorothy	Scientific Name	Present Common Name	Mexican Name
ibis, glossy	*Plegadis chihi*	White-faced Ibis	Atotola
ibis, white	*Eudocimus albus*	White Ibis	Ibis blanco
ibis, wood	*Mycteria americana*	Wood Stork	Cigüeña
jacana	*Jacana spinosa*	Northern Jacana	Cirujano
jay, brilliant (blue back)	*Aphelocoma ultramarina*	Mexican Jay	Grajo azul
jay, brown	*Cyanocorax morio*	Brown Jay	Papán oscuro
jay, magpie	*Calocitta formosa*	White-throated Magpie-Jay	Urraca copetona
killdeer	*Charadrius vociferus*	Killdeer	Tildío
kingbird, Cassin's	*Tyrannus vociferans*	Cassin's Kingbird	Madrugador chilero
kingbird, Couch's	*Tyrannus couchii*	Couch's Kingbird	Madrugador
kingfisher, ringed	*Megaceryle torquata*	Ringed Kingfisher	Pescador gigante
kinglet, ruby-crowned	*Regulus calendula*	Ruby-crowned Kinglet	Reyezuelo de rojo
kite, white-tailed	*Elanus leucurus*	White-tailed Kite	Milano coliblanco
meadowlark	*Sturnella magna*	Eastern Meadowlark	Triguera común
meadowlark	*Sturnella neglecta*	Western Meadowlark	Triguera de occidente
mockingbird	*Mimus polyglottos*	Northern Mockingbird	Centzontle norteño
motmot	*Momotus* sp.	Unknown Motmot	Turco
nonpareil, Leclancher's	*Passerina leclancherii*	Orange-breasted Bunting	Gorrión pecho amarillo
oriole, Alta Mira	*Icterus gularis*	Altamira Oriole	Calandria campera
oriole, Audubon's	*Icterus graduacauda*	Audubon's Oriole	Calandria hierbera
oriole, Baltimore	*Icterus galbula*	Baltimore Oriole	Calandria cañera
oriole, hooded	*Icterus cucullatus*	Hooded Oriole	Calandria zapotera
oriole, orchard	*Icterus spurius*	Orchard Oriole	Calandria café
oropendola	*Psarocolius montezuma*	Montezuma Oropendola	Zacua gigante
osprey	*Pandion haliaetus*	Osprey	Gavilán pescador
owl, barn	*Tyto alba*	Barn Owl	Lechuza común

parakeet	*Aratinga canicularis?*	Orange-fronted Parakeet?	Periquillo común
parakeet, sharp-tailed green	?	Unknown Parakeet	Periquillo
parrot, green	*Bolborhynchus lineola*	Barred Parakeet	Catarina listada
parrot, red-headed	*Amazona viridigenalis*	Red-crowned Parrot	Cotorra cabeza roja
parrot, yellow-headed	*Amazona oratrix*	Yellow-headed Parrot	Cotorra cabeza amarilla
pauraque	*Nyctidromus albicollis*	Common Pauraque	Pochocuate
pelican	*Pelecanus occidentalis*	Brown Pelican	Pelícano moreno
pewee, wood	*Contopus virens*	Eastern Wood-Pewee	Tengofrío verdoso
phainopepla	*Phainopepla nitens*	Phainopepla	Capulinero negro
phoebe	*Sayornis phoebe*	Eastern Phoebe	Papamoscas fibí
phoebe, black	*Sayornis nigricans*	Black Phoebe	Papamoscas negro
pigeon, band-tailed	*Patagioenas fasciata*	Band-tailed Pigeon	Paloma de collar
pigeon, red-billed	*Patagioenas flavirostris*	Red-billed Pigeon	Paloma morada
pintail	*Anas acuta*	Northern Pintail	Pato golondrino
pygmy-owl, ferruginous	*Glaucidium brasilianum*	Ferruginous Pygmy-Owl	Tecolotillo rayado
quail, scaled	*Callipepla squamata*	Scaled Quail	Codorniz escamosa
rail, mangrove	*Aramides axillaris*	Rufous-necked Wood-Rail	Gallineta de collar rojizo
rail, sora	*Porzana carolina*	Sora	Gallineta de ciénega
raven, white-necked	*Corvus cryptoleucus*	Chihuahuan Raven	Cuervo de cuello blanco
redstart, American	*Setophaga ruticilla*	American Redstart	Calandrita
redstart, painted	*Myioborus pictus*	Painted Redstart	Pavito ocotero
roadrunner	*Geococcyx velox*	Lesser Roadrunner	Correcamino veloz
robin, black	*Turdus infuscatus*	Black Robin	Primavera negra
robin, collared	*Turdus rufitorques*	Rufous-collared Robin	Primavera collarejo
sandpiper, least	*Calidris minutilla*	Least Sandpiper	Chichicuilote mínimo
sandpiper, spotted	*Actitis macularius*	Spotted Sandpiper	Alzacolita
scaup	*Aythya affinis*	Lesser Scaup	Pato bola

Name Given by Dorothy	Scientific Name	Present Common Name	Mexican Name
seedeater, Sharpe's	*Sporophila torqueola*	White-collared Seedeater	Collarejito
shoveler	*Anas clypeata*	Northern Shoveler	Pato cucharón
shrike	*Lanius ludovicianus*	Loggerhead Shrike	Verdugo
snipe, Wilson's	*Gallinago delicata*	Wilson's Snipe	Agachona común
solitaire	*Myadestes townsendi*	Townsend's Solitaire	Jilguero común
sparrow	*Passer domesticus*	House Sparrow	Gorrión inglés
sparrow, field	*Spizella pusilla*	Field Sparrow	Chimbito llanero
sparrow, lark	*Chondestes grammacus*	Lark Sparrow	Chindiquito
sparrow, song	*Melospiza melodia*	Song Sparrow	Zorzal cantor
sparrow, Texas	*Arremonops rufivirgatus*	Olive Sparrow	Gorrión oliváceo
spoonbill, roseate	*Platalea ajaja*	Roseate Spoonbill	Cucharera
stilt, black-necked	*Himantopus mexicanus*	Black-necked Stilt	Candelero
swallow, barn	*Hirundo rustica*	Barn Swallow	Golondrina tijerilla
tanager, blue	*Thraupis episcopus*	Blue-gray Tanager	Naranjero nevado
tanager, unknown	*Piranga rubra?*	Summer Tanager	Piranga rubra
tanager, unknown	*Euphonia elegantissima*	Elegant Euphonia	Monjita de capucha azul
teal, blue-winged	*Anas discors*	Blue-winged Teal	Cerceta aliazul
teal, cinnamon	*Anas cyanoptera*	Cinnamon Teal	Cerceta café
teal, green-winged	*Anas crecca*	Green-winged Teal	Cerceta aliverde
tern, least	*Sternula antillarum*	Least Tern	Golondrinita marina
thrasher, Brownsville	*Toxostoma longirostre*	Long-billed Thrasher	Cuitlacoche alacranero
thrasher, curve-billed	*Toxostoma curvirostre*	Curve-billed Thrasher	Cuitlacoche común
tiger-bittern, Mexican	*Tigrisoma mexicanum*	Bare-throated Tiger-Heron	Garza tigre
tinamou	*Crypturellus cinnamomeus*	Thicket Tinamou	Perdiz canela
trogon	*Trogon* sp.	Unknown Trogon	Trogón
turkey	*Meleagris gallopavo*	Wild Turkey	Guajolote
vulture, black	*Coragyps atratus*	Black Vulture	Zopilote común

vulture, turkey	*Cathartes aura*	Turkey Vulture	Aura cabeciroja
warbler, black-throated gray	*Dendroica nigrescens*	Black-throated Gray Warbler	Verdín gargantinegro
warbler, black-throated green	*Dendroica virens*	Black-throated Green Warbler	Verdín de pecho negro
warbler, little yellow	*Geothlypis* sp.?	Unknown Warbler	
warbler, myrtle	*Dendroica coronata*	Yellow-rumped Warbler	Verdín de toca
warbler, Nashville	*Vermivora ruficapilla*	Nashville Warbler	Verdín de mono
warbler, olive	*Peucedramus taeniatus*	Olive Warbler	Verdín oliváceo
warbler, orange-crowned	*Vermivora celata*	Orange-crowned Warbler	Gusanero cabecigrís
warbler, pileolated	*Wilsonia pusilla*	Wilson's Warbler	Pelucilla
warbler, Tennessee	*Vermivora peregrina*	Tennessee Warbler	Verdín semillero
warbler, Wilson's	*Wilsonia pusilla*	Wilson's Warbler	Pelucilla
waxwing, cedar	*Bombycilla cedrorum*	Cedar Waxwing	Chinito
willet	*Tringa semipalmata*	Willet	Zarapico semipalmeado
woodhewer, ivory-billed	*Campephilus guatemalensis*	Pale-billed Woodpecker	Picotero
woodpecker, acorn-eating	*Melanerpes formicivorus*	Acorn Woodpecker	Carpintero encinero
woodpecker, brown	*Picoides stricklandi*	Strickland's Woodpecker	Carpintero de strickland
woodpecker, ladder-backed	*Picoides scalaris*	Ladder-backed Woodpecker	Carpintero listado
woodpecker, red-headed	*Campephilus imperialis*	Imperial Woodpecker	Pitorreal ocotero
woodpecker, golden-fronted	*Melanerpes aurifrons*	Golden-fronted Woodpecker	Cheque
wren, cactus	*Campylorhynchus brunneicapillus*	Cactus Wren	Matraca grande
wren, canyon	*Catherpes mexicanus*	Canyon Wren	Saltapared risquero
wren, house	*Troglodytes aedon*	House Wren	Saltapared cucarachero
wren, rock	*Salpinctes obsoletus*	Rock Wren	Saltaladera
yellowthroat, marsh or Maryland	*Geothlypis trichas*	Common Yellowthroat	Tapaojito común

American Ornithologists' Union (1998, 2000, 2002, 2003, 2004, 2005, 2006, 2007); Anonymous (2002); Birkenstein and Tomlinson (1981); Peterson (1941); Peterson and Chaliff (1973); Sibley (2000).

Notes

Preface

1. Also, it should be noted that in addition to the Saunderses' ground survey, a second U.S. team composed of Flyway Biologist–Pilot Robert H. Smith and U.S. Game Management Agent Floyd Thompson conducted the February 1949 aerial survey of waterfowl wintering along the Gulf Coast and Caribbean, while Smith and Flyway Biologist–Pilot David L. Spencer by plane later inventoried populations wintering along the Mexican Pacific coast (see Saunders and Saunders 1981, table 1). No coverage was made that winter of wetlands and waterfowl in the northern interior highlands of Mexico (Rod C. Drewien, pers. comm., December 5, 2005).

Introduction

1. Her dissertation as well as portions thereof were published as Saunders [Carpenter] (1937, 1938).

2. Jones (1997).

3. Saunders, Handley, and Holloway (1950).

4. An armada was an arrangement of many, often dozens, of side-by-side gun barrels set into the earth in a semicircle, partially surrounding a shallow-water waterfowl feeding area, often baited for several prior days. When a large number of waterfowl had been attracted to and concentrated in the area, the guns were fired simultaneously into the mass by means of fuses, from a remote location. The barrels were angled slightly to cover a wide width of the feeding birds upon discharge; thus, the resulting kills were often huge, numbering into the many hundreds if not thousands. Concealed gatherers then rushed forth to retrieve and kill the cripples. Such large kills were usually sold in nearby cities. The armada was among the deadliest means ever devised by humans to kill enormous numbers of waterfowl. Their use in North America was prohibited long ago. See Pearson (1937).

5. Frenetic diplomatic efforts in Mexico sought to conclude the treaty for announcement at the first North American Wildlife Conference, called by President Franklin D. Roosevelt, and held in Washington, D.C., February 3–7, 1936. Juan Zinser had the honor to proclaim at the Friday afternoon concluding session: "Now I am going to announce to you that the treaty between the United States of America and Mexico has been signed provisionally this afternoon. [Applause]. This treaty will be signed this evening

at 6 o'clock, Mexican time, definite. [Applause]" (Zinser 1936c, 254). Zinser was chief of the newly created Game Section, Department of Forestry, Game, and Fish, Mexico City. E. A. Goldman and his son, L. C., describe the meetings and final negotiations during January 10 through February 7, 1936 (Goldman and Goldman ca. 1936).

6. The Saunderses, following the more basic inventories of Edward W. Nelson and the Goldmans, Edward A., Luther J., and Luther C., continued this work. George Saunders is credited with making the first aerial waterfowl wintering ground survey in 1938 (Saunders and Saunders 1981). The same source lists participants during the 1939–64 period (see also Hawkins et al. 1984).

7. People working for the U.S. government and traveling to foreign countries are required to carry official passports rather than regular passports, and many Mexican border immigration officials do not know how to process them. This invariably leads to delays in obtaining visas. Similarly, problems are generated by U.S. government vehicles. In addition, anyone bringing firearms into the country must visit the local Campo Militar to obtain gun permits, perhaps a bit easier in 1948 than today (each port of entry has a military post that, among its other duties, handles requests to bring arms in the country). However, with a little *mordida* (the bite) to one or more inspectors, the way was usually greased considerably and one could avoid extensive luggage inspection and be on the way a little sooner.

8. Even U.S. military maps that they carried sometimes were found to be unreliable.

9. Most fields and pastures were unfenced, and livestock frequently wandered onto roads, much like today in "open ranges" in our own western states.

10. In the 1940s the water supplies in most towns were subject to contamination. Many restaurant diners drank beer and soft drinks and never ate raw fruits and vegetables during their meals to avoid a case of so-called Montezuma's Revenge. Every precaution had to be taken. Such hazards help explain the Saunderses' reliance upon canned foods and peanut butter sandwiches. Today, these problems have largely been removed.

11. Pulque is one of several Mexican alcoholic beverages fermented and distilled from the juice of the agave or maguey plant. Tequila is a higher-quality distillate.

Chapter 1

1. In the field trip technical report, the Saunderses (1950) list their objectives during the January–May 1949 investigations: "(1) take the January waterfowl inventory at lakes of the interior plateau from Guanajuato and Jalisco south to the Valley of Mexico; (2) study general conditions at several important wintering grounds; (3) establish contacts with officials and hunters in Mexico, D.F.; (4) contact hunters in several other important localities to obtain local information on waterfowl, as well as their assistance in conducting their own field investigations; (5) obtain information on other species of migratory birds besides waterfowl, particularly mourning doves and white-winged doves." They also were to collect birds for the U.S. National Museum, Washington, D.C.

2. Dorothy identifies herself on the technical report (Saunders and Saunders 1950) simply as "Collaborator." Perhaps by this time she recognized that her assistance to George was far broader than being a "botanical collaborator."

3. In several locations throughout Dorothy's account she mentions visiting markets or *mercados*. Nearly all cities and towns of any size in Mexico have a centralized *mercado*, or general store. Typically, they have two stories, open in the middle with individual shops located on balconies all around the perimeter inside and sometimes outside the building. Meats, vegetables, and fruits are normally sold on the ground floor in the middle by merchants. The perimeter shops sell everything imaginable, from legal drugs, to hats, to leather goods, to toys. Anyone who likes to shop finds the local *mercado* irresistible, and a little bargaining usually results in a purchase satisfying to both the shop owner and the buyer. Dorothy obviously enjoyed visiting each *mercado* when she had the opportunity and frequently made purchases.

4. Dorothy repeatedly and indiscriminately refers to many rural people they encountered as "Indios." This term conjures up an image of primitive people still living in the Stone Age. In fact, many of the people they encountered were probably "mestizos," or of mixed blood among the Spanish and Indian citizens that originated up to over four centuries ago at the Conquest. Although there were still many pure-blooded "Indians" or Native Americans in remote places of Mexico in the 1940s, most people were of mixed ancestry living in domestic, although Spartan, comfort. So we advise the readers that the Indios of Dorothy's account were not always "Indians" as we might envision them. To soften Dorothy's overly repetitive use of "Indios," we sometimes have substituted "natives," "campesinos," or other euphemisms.

5. George Saunders, when disposing of some of his possessions, gave his altimeter to Paul Springer, who in turn donated it to the historical collection of the U.S. Fish and Wildlife Service's National Conservation Training Center, Shepherdstown, West Virginia.

6. The Aztecs built several causeways across Lake Texcoco to their capital, Tenochtitlán.

7. "Established temporary office and laboratory facilities at the headquarters of the U.S. Bureau of Entomology and Plant Quarantine, located in the San Jacinto section of the city, adjacent to the Mexican Department of Agriculture buildings" (Saunders and Saunders 1950, 9).

8. An intended pun?

9. Late in 1948, Albert M. Day, director of the Fish and Wildlife Service, along with staff members Logan J. Bennett, chief of the Branch of Research, and David R. Gascoyne, assistant chief of the Branch of Game Management, visited Mexico to view the duck situation firsthand (Day 1949, 265–85). The trip was prompted by a widespread belief among American hunters that Mexicans were slaughtering and selling enormous numbers of ducks while the U.S. harvest was being curtailed by very restrictive hunting regulations. The U.S. contingent met with Mexican officials and hunters and visited several lake remnants in the Valley of Mexico, including Lake Texcoco. Although they did find a few Mexicans hunting with ancient firearms, saw the kill resulting from one armada firing (subsequently confiscated by Mexican officials), and occasionally saw a few ducks being sold on the main streets of Mexico City, they concluded that the Mexican kill was insignificant. They estimated that only about a thousand Mexicans purchase hunting licenses for waterfowl hunting (compared with 1.75 to 2.0 million in the United States) and that the total Mexican kill did not exceed 5 percent of the total North American kill.

10. "The shallow basin which formerly contained broad Lake Texcoco now has relatively little open water. Drainage from the lake and diversion of inflow have changed it from one of the best waterfowl wintering grounds in the interior to one of steadily dwindling value. . . . At the present time the basin is divided into units by dikes and controlled by the neighboring villages" (Saunders and Saunders 1950, 89–90).

11. Aztec hydrologists had constructed many types of water-control structures in the Valley of Mexico during the pre-Conquest period. These included numerous canals; aqueducts; dikes to divide the waters of Lake Texcoco; and wetlands filling to create desperately needed farmland, *chinampas,* upon which to produce crops for the burgeoning population; and causeways for transportation to their water-surrounded capital, Tenochtitlán. Some preliminary work commenced on the drainage of the various lakes to reduce flooding and create even more farmland. In 1607, the Spanish viceroy approved a plan developed by Enrico Martínez to begin drainage of the northern lakes by means of canals, and a tunnel near the present town of Tequixquiac. This continued the long process of converting the closed basin to a valley having drainage to the Gulf of Mexico. This initial General Drain, the *Desagüe General,* was completed in less than a year by some sixty thousand Indian laborers at the expenditure of a fifth of the entire colonial budget (Simon 1997). The work proved inadequate, as did a series of other drainage efforts that continued intermittently over the following nearly three centuries. President Diaz hailed the completion of the project as a symbol of Mexico's entry into the modern world. Conversely, Simon stated, "The draining of Lake Texcoco marked one of the most monumental ecological transformations in human history" (1997, 68).

12. The Saunderses were aware of Director Day's visit to Mexico, for they state, "This [La Cueva, a diked unit in Lake Texcoco] is one of the places which Director Albert M. Day, Dr. Logan Bennett and R. Gascoyne visited on their trip last December" (Saunders and Saunders 1950, 90).

13. "At time of the January inventory we were unable to gain admittance to two of the best units, possibly because an armada is reported to be located there and shot once a week" (Saunders and Saunders 1950, 90).

In his opening remarks at the Washington, D.C., conference, Juan Zinser (1936a) noted that the use of armadas had been forbidden four years earlier by presidential decree but that their use was occasionally practiced by evading official vigilance. In 1938, Zinser reported that despite strong enforcement action, use of the armada continued. In one confrontation when federal officers sought to confiscate ducks shot by the firing of an armada, several peasants where killed or wounded. In 1944, Zinser estimated that despite federal enforcement, an average of two armada firings a day occurred in the Valley of Mexico and the Río Lerma, with each discharge killing six hundred to seven hundred ducks. In 1949, Luis Macias Arellano reported that commercial waterfowl hunting was practiced chiefly to supply markets and restaurants in Mexico City and along the U.S. border catering to foreign tourists and visitors.

14. In the acknowledgments to their official report the Saunderses (1950) express appreciation to Luis Macias and Manuel J. Guerrero of the Servicio de Caza. The latter is no doubt "Sr. G." to whom Dorothy refers.

15. During January 1–3, 1949, the distribution of ducks in the Valley of Mexico was as follows: Lake Texcoco, 15,000; Lake Tultengo, 4,500; Lake Zumpango, 500; and other smaller lakes, 10,000 (Saunders and Saunders 1950, 135).

Chapter 2

1. "Lake Cuitzeo is located . . . fifteen miles north of Morelia at an elevation of about 6,000 feet. It is approximately thirty-two miles long and is up to six miles wide. Extensive marshes occur at the delta of the Río Morelia and concentrations of ducks and several thousand white-fronted geese were observed in that vicinity" (Saunders and Saunders 1950, 86).

2. The 1983 American Ornithologists' Union's *Check-list* withdrew recognition of the Mexican duck (*Anas diazi* Ridgway) as a distinct species and designated it as simply a form of the mallard (*A. platyrhynchos* Linnaeus).

3. "Lake Pátzcuaro, located twenty five miles east southeast of Morelia, at an elevation of 7,182 feet, is one of the best duck lakes in the interior of Mexico. It measures approximately fifteen miles in length with its width varying from about one and a half miles up to four or five miles" (Saunders and Saunders 1950, 87).

4. In the January 1949 aerial survey, 62,000 ducks and 76,000 coots were present on Lake Pátzcuaro (Saunders and Saunders 1950, 130).

Chapter 3

1. A recent news article reports the continuing decline of Lake Chapala, Mexico's largest lake, regarded as a "national treasure" (Carlton 2003). Since the 1970s the lake has lost about 80 percent of its water due to heavy development, particularly the diversion of water from the Río Lerma for irrigation. Some ten thousand U.S. and Canadian expatriates, who are estimated to contribute $200 million to the local economy, now dwell along its former shoreline far from water. Meanwhile, the lake's former fish production has plummeted because of lowering water levels, sewage, and other pollution. The whitefish, famed for its delicate taste and regarded by the Saunderses as an epicurean delight, has essentially vanished. Nonetheless, the National Water Commission plans to restore the lake to former levels by 2010. Critics are highly skeptical.

2. The Saunderses estimated, based on the January 1949 aerial survey, that Lake Chapala, excluding the delta marshes and adjoining fields, held approximately 38,000 ducks and 18,000 coots (Saunders and Saunders 1950).

3. The Saunderses (1950) noted that Lake Chapala, measuring about forty-three miles long and twelve wide at its broadest width, is the largest lake in Mexico. Waterfowl usage was concentrated in the rich marshes of the Río Lerma delta. "Considering the value of this lake to waterfowl and the large numbers present, hunting pressure is amazingly light. The city of Guadalajara is approximately 25 miles from the nearest point on the lakeshore and probably about twice that distance from the delta, but few hunters seem to avail themselves of the shooting opportunities there" (Saunders and Saunders 1950, 85–86).

4. George had studied meadowlarks in earlier graduate studies and continued to be actively interested in their taxonomy. As with white-winged doves, he proposed several

races or subspecies that were never officially adopted. At the present time, two species are recognized by the AOU: eastern (*Sturnella magna*) and western (*S. neglecta*) meadowlarks.

5. "Tortilla con chile" may have been the way the meadowlark's song sounded to the listener, as "Who cooks for you?" onomatopoetically describes the song of the white-winged dove.

Chapter 4

1. Extrapolating from a 25 percent aerial coverage of the vast lagoons near Mazatlán on January 19, 1948, the Saunderses estimated that approximately 1.1 million ducks were present. They remarked that the area is one of the outstanding waterfowl wintering areas remaining in Mexico (Saunders and Saunders 1950, 71).

2. As far as the editors know, this race was never formally considered or adopted by the AOU.

3. George published an article (Saunders 1968) in which he described twelve subspecies of white-winged doves in North, Central, and South America, four of which breed in the United States. The eastern (*Zenaida asiatica asiatica*) and the western (*Z. a. mearnsii*) are the most prominent populations; management decisions by the FWS are made separately each year for each.

Chapter 5

1. During the three days spent scouting the lake in January, from eight to fourteen dugout canoes, each with two Tarascan Indians, were engaged in duck and coot hunting using old black powder, muzzle-loading shotguns and atlatls. Their harvest ratio was approximately 20 coots to every duck. On market day, Friday, January 27, 1948, approximately 250 coots and 15 ducks, mostly ruddy ducks and cinnamon teal, were offered for sale (Saunders and Saunders 1950, 88).

Chapter 6

1. On the January 15, 1948, aerial survey an estimated 164,000 ducks and 6,200 coots were on the Laguna Papagallo or Tres Palos Lagoon. Foremost in numbers was the baldpate, followed by the pintail and shoveler, and gadwall (Saunders and Saunders 1950).

Chapter 8

1. In the late 1970s, while on an official wildlife trip to Venezuela (Reeves), our team stopped in a small, northern Andean town. We observed that residents were evacuating their homes and assembling in the street in a semi-gala mood. A large tank truck with a noisy pump mounted on the rear appeared, and workers carried a flexible pipe with nozzle indoors. Soon a whitish fog was escaping from all windows and doors. We were told that it was the "día de DDT." Apparently the Saunderses had no ill effect from any DDT encounters as both lived into their nineties.

2. The immense Tule cypress at Santa María del Tule, Oaxaca, still survives. Studies using DNA techniques document that it is indeed a single tree, not two as some have

suggested. The tree has a circumference of 164 feet (54 meters), thought to be the greatest girth of any tree on earth. Its weight has been estimated at 550 tons and its volume at 705 cubic meters. It is thought to be between two thousand and three thousand years old (George DeLange and Audrey DeLange, *Tule tree* [Oaxaca: Santa María del Tule, 2005], an eight-page booklet of travel photos).

Chapter 9

1. Cabrito, a young kid (goat), usually barbecued, and served with a sauce, is a noted delicacy of northern Mexico. Because it is sometimes greasy, other meat dishes such as carne asada, a marinated and grilled beef tenderloin, are preferred by some Americans. Both editors find either acceptable.

2. The only long-tailed jays in Mexico are two species of magpie jays, but they both occur south of San Luis Potosí.

3. We wonder whether while gazing upon the Río Pánuco from their hotel lawn chairs, the Saunderses realized that it was the headwaters of this very river into which the waters of the magnificent lakes and marshes of the great Valley of Mexico were drained into the Gulf of Mexico. The vehicle for the successive drainage of these wetlands by the Aztecs, Spaniards, Mexicans, and English entrepreneurs and engineers, now regarded as one of the greatest ecological disasters in North America, was closely before them.

4. Dorothy later studied sharks at the Cape Haze Marine Laboratory, Sarasota, Florida, and no doubt realized that Beebe was mistaken.

5. The 1948 survey was conducted with Flyway Biologist David L. Spencer (Saunders and Saunders 1981).

6. In 1949, the illegal drug business had not yet become reality. Coeditor Tomlinson, in later years, witnessed several instances where *contrabandistas* were operating and he and his cohorts had to be especially circumspect to avoid confrontations with them.

7. The first Roger Tory Peterson field guide, for eastern birds, appeared in 1934; that for western birds in 1941; and for birds of Mexico, 1973.

Chapter 10

1. "I would judge from all the information received from these duck hunters, American and Mexican, that there are fewer than fifty sportsmen in that city of more than 2,000,000 people who hunt ducks more or less regularly. Their 1948–49 season's bag was probably less than 5,000 ducks" (Saunders and Saunders 1950, 91).

2. It was this very market that the Goldmans visited on January 9 and 19, 1936. On the former date they reported: "Found a total of about 150 ducks on sale at three stands. About 100 of these were pintails and about 20 spoonbills [Northern Shovelers], and the remainder about equally divided between blue-winged teal, cinnamon teal, and green-winged teal. The price asked for the pintails was $1.75 per pair. The teals were offered for sale somewhat cheaper owing to their smaller size" (Goldman and Goldman ca. 1936, 2). At the first North American Wildlife Conference, Juan Zinser stated, "By decree in 1936 the selling of wild ducks is absolutely prohibited" (1936b, 120). However, he also observed the

great difficulties in enforcing game laws over such a large region as Mexico. We might add, in the United States as well.

3. Sale of captured wild birds was conducted openly throughout Mexico until the 1960s. Since then, the practice has continued furtively, despite enforcement efforts to stop it. Because of this widespread, illegal taking of birds, parrot and parakeet populations have declined substantially in many parts of the country.

References

Mentioned in Text, Notes, or Captions

American Ornithologists' Union. 1983. *Check-list of North American birds.* 6th ed. Washington, D.C.: American Ornithologists' Union.

———. 1998. Check-list of North American birds. 7th ed. Washington, D.C.: American Ornithologists' Union.

———. 2000. Check-list of North American birds. 42nd supplement. *The Auk* 117: 847–58.

———. 2002. Check-list of North American birds. 43rd supplement. *The Auk* 119: 897–906.

———. 2003. Check-list of North American birds. 44th supplement. *The Auk* 120: 923–32.

———. 2004. Check-list of North American birds. 45th supplement. *The Auk* 121: 985–95.

———. 2005. Check-list of North American birds. 46th supplement. *The Auk* 122: 1026–31.

———. 2006. Check-list of North American birds. 47th supplement. *The Auk* 123: 926–36.

———. 2007. Check-list of North American birds. 48th supplement. *The Auk* 124: 1109–15.

Anonymous. 2002. *Field guide to the birds of North America.* 4th ed. Washington, D.C.: National Geographic Society.

Arellano, Luis Macias. 1949. Wildlife problems in Mexico. *Transactions of the North American Wildlife Conference* 14:9–16.

Carlton, Jim. 2003. A lake shrinks, threatening Mexican region. *Wall Street Journal,* September 3, 2003, B1.

Day, Albert M. 1949. *North American waterfowl.* Harrisburg, Penn.: Stackpole.

Godman, Frederick D., and Osbert Salvin. 1879–1915. *Biologia Centrali-Americana.* 63 vols. London: R. H. Porter.

Goldman, E. A., and L. C. Goldman. ca. 1936. Wintering of migratory waterfowl in Mexico, January 5 to May 23, 1936. Typed report. Washington, D.C.: Bureau of Biological Survey.

Hawkins, A. S., R. C. Hanson, H. K. Nelson, and H. M. Reeves. 1984. *Flyways, pioneering waterfowl management in North America.* Washington, D.C.: Fish and Wildlife Service, Government Printing Office.

Jones, Elliott. 1997. A life full of memories. *Press Journal,* Vero Beach, Florida, August 10, 1997.

Meyer, Michael C., and William H. Beezley. 2000. *The Oxford history of Mexico.* New York: Oxford University Press.

Pearson, T. Gilbert. 1937. *Adventures in bird protection.* New York: D. Appleton–Century.

Peterson, R. T. 1941. *A field guide to Western birds.* Cambridge, Mass.: Riverside Press.

Peterson, R. T., and E. L. Chaliff. 1973. *Mexican birds.* Norwalk, Conn.: Easton Press.

Reeves, Henry M. 2001. In memoriam: George B. Saunders. *Auk* 118 (4): 1030–31.

Roetker, F. H., and J. S. Wortham. 2006. *Winter waterfowl survey, east coast of Mexico: Rio Grande Delta to northeastern Yucatan.* Lafayette, La.: U.S. Fish and Wildlife Service.

Saunders [Carpenter], Dorothy. 1937. Anatomy of the leaf transition region of *Urginea maritima. Papers of the Michigan Academy of Science, Arts, and Letters* 22:27–31.

———. 1938. Anatomy of the inflorescence of *Urginea maritima* (L.) Baker. *Papers of the Michigan Academy of Science, Arts, and Letters* 23:109–15.

Saunders, Dorothy Chapman. 1951. Territorial songs of the White-winged Dove. *Wilson Bulletin* 63 (4): 330–32.

Saunders, George B. 1952. Waterfowl wintering grounds in Mexico. *Transactions of the North American Wildlife Conference* 17:89–100.

———. 1964. South of the border. In *Waterfowl tomorrow,* edited by Joseph P. Linduska, 253–62. Washington, D.C.: USDI, Fish and Wildlife Service, U.S. Government Printing Office.

———. 1968. *Seven new subspecies of White-winged Doves from Mexico, Central America and southwestern United States.* North American Fauna No. 65. Washington, D.C.: U.S. Fish and Wildlife Service, Bureau of Sport Fisheries and Wildlife.

———. 1984. The U.S. team. In *Flyways,* edited by A. S. Hawkins, R. C. Hanson, H. K. Nelson, and H. M. Reeves, 121–29. Washington, D.C.: Fish and Wildlife Service.

Saunders, George B., C. O. Handley, and Ancil D. Holloway. 1950. *A fish and wildlife survey of Guatemala.* U.S. Fish and Wildlife Service, Special Scientific Report No. 5. Washington, D.C.: Government Printing Office.

Saunders, George B., and Dorothy C. Saunders. 1950. Report on migratory waterfowl investigations in Mexico, January–May, 1949. U.S. Fish and Wildlife Service, Washington, D.C. Typed mss. Receipt stamped, March 15, 1950.

———. 1981. *Waterfowl and their wintering grounds in Mexico, 1937–64.* Resource Publication 138. Washington, D.C.: U.S. Fish and Wildlife Service.

Saunders, George B., P. B. Uzzell, S. Gallizioli, W. H. Kiel Jr., J. A. Neff, and J. Stair. 1968. Ed. Clarence Cottam and James B. Trefethen. *Whitewings, the life history, status, and management of the White-winged Dove.* Princeton, N.J.: Van Nostrand.

Secretaría de Medio Ambiente y Recursos Naturales. 2002. Ley General de Vida Silvestre. México, D. F., México.

Sibley, D. A. 2000. *National Audubon Society The Sibley guide to birds.* New York: Alfred A. Knopf.

Simon, Joel. 1997. *Endangered Mexico, an environment on the edge.* San Francisco: Sierra Club Books.

Thorpe, P., R. Drewien, and G. Ojeda García. 2006. *Winter waterfowl survey, Mexico: interior highlands and lower west coast.* Denver, Colo.: U.S. Fish and Wildlife Service.

Valdez, R., J. C. Guzmán-Aranda, F. J. Abarca, L. A. Tarango-Arámbula, and F. Clemente-Sánchez. 2006. Wildlife Conservation and Management in Mexico. *Wildlife Society Bulletin* 34(2): 270–82.

Zinser, Juan. 1936a. Remarks. *Transactions of the North American Wildlife Conference* 1:6–11.

———. 1936b. [Further] Remarks. *Transactions of the North American Wildlife Conference* 1:119–20.

———. 1936c. Address of Señor Juan Zinser, of Mexico. *Transactions of the North American Wildlife Conference* 1:254–55.

———. 1938. What is the status of waterfowl? *Transactions of the North American Wildlife Conference* 3:164–66.

———. 1944. Mexico's conservation program. *Transactions of the North American Wildlife Conference* 9:29–33.

Further Reading

Annerino, John. 1994. *The wild country of Mexico.* San Francisco: Sierra Club Books.

Arellano, M., and P. Rojas M. 1956. *Aves aquaticas migratorias en Mexico.* Mexico, D.F.: Inst. Mex. Recursos Naturales Renovables.

Birkenstein, Lillian R., and Roy E. Tomlinson. 1981. *Native names of Mexican birds.* Resource Publication 139. Washington, D.C.: U.S. Fish and Wildlife Service.

Blake, Emmet R. 1953. *Birds of Mexico: A guide for field identification.* Chicago: University of Chicago Press.

Burkholder, Mark A., and Lyman L. Johnson. 2001. *Colonial Latin America.* 4th ed. New York: Oxford University Press.

Coe, Michael D. 1962. *Mexico, ancient peoples and places.* New York: Frederick A. Praeger.

Cortés, Hernán. 1986. Trans. and ed. Anthony Pagden. *Letters from Mexico.* New Haven, Conn.: Yale University Press.

Davis, Irby. 1972. *A field guide to the birds of Mexico and Central America.* Austin: University of Texas Press.

Díaz del Castillo, Bernal. 1974. Trans. J. M. Cohen. *The conquest of New Spain.* London: Folio Society.

Dillehay, Thomas D. 2000. *The settlement of the Americas.* New York: Basic Books.

Edwards, Ernest P. 1968. *Finding birds in Mexico.* Sweet Brian, Va.: Ernest P. Edwards.

———. 1998. *A field guide to the birds of Mexico and adjacent areas: Belize, Guatemala, and El Salvador.* Austin: University of Texas Press.

Eisenmann, Eugene. 1955. *The species of Middle American birds.* Transactions, vol. 7. New York: Linnaean Society of New York.

Exequiel, Excurra, Marisa Mazari-Hiriat, Irene Pisanty, and Adrián Guillermo Aguilar. 1999. *The basin of Mexico: Critical environmental issues and sustainability.* New York: United Nations University Press.

Foster, George M. 1948. *Empire's children, the people of Tzintzuntzan.* Institute of Social Anthropology Publication No. 6. Washington, D.C.: Smithsonian Institution; Mexico: Imprenta Nuevo Mundo, S.A.

Friedmann, Herbert, Ludlow Griscom, Robert T. Moore, and Alden H. Miller. 1950, 1957. *Distributional check-list of the birds of Mexico. Parts 1 and 2.* Pacific Coast Avifauna, Nos. 29 and 33. Berkeley: Cooper Ornithological Club.

Gadow, Hans. 1908. *Through southern Mexico, being an account of the travels of a naturalist.* London: Witherby.

George, Ronnie R., Roy E. Tomlinson, Ronald W. Engel-Wilson, Gary Waggerman, and A. Gordon Spratt. 1994. White-winged Dove. In *Migratory shore and upland game bird management in North America,* edited by Thomas C. Tacha and Clait E. Braun, 28–50. Washington, D.C.: International Association of Fish and Wildlife Agencies.

Howell, Steve N. G., and Sophie Webb. 2004. *A guide to the birds of Mexico and northern Central America.* New York: Oxford University Press.

Kramer, Gary W., Eduardo Carrera, and Daniel Zaveleta. 1995. Waterfowl harvest and hunting activity in Mexico. *Transactions of the 60th North American Wildlife and Natural Resources Conference* 60:243–50.

Kramer, Gary W., and Rodrigo Migoya. 1989. The Pacific coast of Mexico. In *Habitat management for migratory waterfowl in North America,* edited by Loren M. Smith, Roger L. Pederson, and Richard M. Kaminski, 507–28. Lubbock: Texas Tech University Press.

Leopold, A. Starker. 1959. *Wildlife of Mexico, the game birds and mammals.* Berkeley: University of California Press.

———. 1964. Mexico and migratory waterfowl conservation. In *Waterfowl tomorrow,* edited by Joseph P. Linduska, 729–36. Washington, D.C.: Fish and Wildlife Service.

Nelson, E. W. 1921. *Lower California and its natural resources.* National Academy of Sciences, 16, First Memoir. Washington, D.C.

Niederberger, Christine. 1979. Early sedentary economy in the basin of Mexico. *Science* 203 (4376): 131–42.

Nuttall, Zelia. 1891. *The atlatl or spear-thrower of the ancient Mexicans.* Paper of the Peabody Museum of Archaeology and Ethnology, vol. 1, no. 3. Cambridge, Mass.: Harvard University.

Peterson, Roger Tory, and Edward L. Chaliff. 1973. *A field guide to Mexican birds and adjacent Central America.* Boston: Houghton Mifflin.

Prescott, William H. 1843, 1931. *History of the conquest of Mexico and history of the conquest of Peru.* New York: Modern Library.

Sabloff, Jeremy A. 1989. *The cities of ancient Mexico, reconstructing a lost world.* New York: Thames and Hudson.

Salvin, O., and F. D. Godman. 1879–1904. *Biologia Centrali-Americana.* Vol. 3, *Aves.* London: R. H. Porter.

Sanders, William T., Jeffrey R. Parsons, and Robert S. Santley. 1979. *The basin of Mexico, ecological processes in the evolution of a civilization.* New York: Academic Press.

Schwertner, T. W., H. A. Mathewson, J. A. Roberson Jr., M. Small, and G. L. Waggerman. 2002. *White-winged Dove.* The Birds of North America 710, edited by A. Poole and

F. Gill. Philadelphia: Academy of Sciences; Washington, D.C.: American Ornithologists' Union.

Simon, Joel. 1997. Prophecy, plague, and plunder. *The Amicus Journal* 19 (1): 28–33.

Stirling, M. W. 1960. The use of the atlatl on Lake Patzcuaro, Michoacan. Anthropological Papers, No. 59. *Bureau of American Ethnology Bulletin* 173:261–68.

Sutton, G. M. 1951. *Mexican birds. First impressions.* Norman: University of Oklahoma Press.

———. 1972. *A bend in a Mexican river.* New York: Paul S. Erickson.

Thomas, Hugh. 1993. *Conquest. Montezuma, Cortés, and the fall of old Mexico.* New York: Touchstone, Simon and Schuster.

Townsend, Richard F. 1992. *The Aztecs.* New York: Thames and Hudson.

Wauer, Roland H. 1999. *Birder's Mexico.* College Station: Texas A&M University Press.

West, Rebecca. 2003. Ed. Bernard Schweizer. *Survivors in Mexico.* New Haven, Conn.: Yale University Press.

Wheatley, Nigel, and David Brewer. 2002. *Where to watch birds in Central America, Mexico, and the Caribbean.* Princeton, N.J.: Princeton University Press.

Wolf, Eric R., ed. 1976. *The valley of Mexico, studies in pre-Hispanic ecology and society.* Albuquerque: School of American Research and the University of New Mexico Press.

Wood, Michael. 2000. *Conquistadors.* Berkeley: University of California Press.

Index

ISBN-13: 978-1-60344-061-5
ISBN-10: 1-60344-061-5
53400
9 781603 440615